PREPPER SUPPLIES & SURVIVAL GUIDE

PREPPER SUPPLIES
& SURVIVAL GUIDE

THE PREPPING **SUPPLIES, GEAR, & FOOD** YOU MUST HAVE TO SURVIVE

NOVATO
PRESS

CONTENTS

Preparing House and Mind

W hy bother being prepared for a disaster? Typically, we Americans navigate our daily routines as smoothly as we drive our cars. But much of our confidence comes from the nature of routine itself, from living day to day and year to year in fairly predictable circumstances—intact homes, functioning hospitals, running water, stocked supermarkets that open on schedule, and everyone *where* they should be *when* they should be there. It's natural to take for granted that we have things under control.

But when disaster strikes—as anyone who's lived through it knows—the fantasy of control collapses, along with the infrastructure that makes our lives normal. When the unthinkable occurs, our perceptions change along with our surroundings; we become viscerally aware that we can't control outside events—and that to stay alive we must act quickly in unfamiliar circumstances.

Thankfully, we *can* survive disasters. History demonstrates that in the face of catastrophic threats, we humans instinctively act to preserve life—both our own and the lives of others. To do so effectively, however, we need the right tools. This book is your guide to those tools, not the least of which are mental tools.

Mental Supplies

Surprisingly—especially for those looking to simply stock up on survival supplies and gear—the most important component to readiness is a prepared state of mind. This includes an understanding that disaster asks us not only to survive, but to help others survive. Naturally, we think first of our loved ones in a disaster. But disaster strikes on its own schedule, when we are at an airport in an unfamiliar city, or on the transit system going home from work, or even in the aisles of the local grocery store. At the moment of crisis, your life may be in the hands of a complete stranger. Or a stranger's survival may depend on you knowing what to do in that moment. Would you know what to do? Are you prepared with the knowledge and skills to save a life or to calm a highly traumatized person?

As you read through this book for the practical (and here and there amusingly excessive examples of) supplies available for disaster preparation, keep in mind that the most valuable thing to have in advance of an emergency is first-aid and Community Emergency Response Team (CERT) training—and a mindset that has moved beyond asking "How will I get my needs met?" to "How will I help others?" Merely stocking survival water, food, and first aid supplies makes you a person with resources, but earning certification in first aid and CERT makes you a resource to your loved ones, coworkers, and community members.

Examples of practical heroism abound during disasters. Don't be misled by the numerous survivalist websites depicting preparedness as a solo endeavor accomplished in grim isolation. The overwhelming majority of Americans live in highly populated areas where disaster relief is a shared effort and preparedness is fast becoming a civic duty. As more and more Americans volunteer to train for emergency preparedness, our country becomes a team of prepared citizens. When disaster strikes, local teams jump into action first; federal disaster management teams quickly follow, staffed by people trained to coordinate response efforts. In a disaster, those with CERT training are identified and given special assignments, since they are known to emergency responders as being knowledgeable and well prepared.

This sorting of volunteers in a disaster is both organizational and life-saving. For example, if you have not been trained to read and understand hazardous materials placards on buildings, you could decide impulsively to rush in and offer help inside a structure that is about to explode. Certain highly dangerous rescue operations must be done by emergency professionals and not the rest of us. A CERT-trained person knows what should not be done as well as what should be done during a disaster. And that knowledge enables those around him or her to sidestep panic and fear.

What Would Mrs. Miniver Do?

We all know that the United States has been and will continue to be a target for terrorist organizations. But just because a terrorist attack is possible, it does not mean it is likely in your neighborhood. Whenever we cultivate fear or panic in ourselves or in others, we are essentially bowing to terrorists, whose main goal is to elevate fear, and through that fearfulness exert control. It is a highly psychological weapon, for which we prepare as a people—not as isolated individuals—by maintaining the state of mind that we can and will survive, all together.

The American capacity for teamwork and positivity during disasters is well established, and has been depicted in countless stories, films, and television shows—showing real events and fictional ones. A classic of the genre is *Mrs. Miniver*, a 1942 film set in wartime Britain but made by an American cast and crew and directed by the legendary William Wyler. Mrs. Miniver is a mild-mannered British housewife who conducts herself with extraordinary level-headedness and courage even when under attack. Recognized by the American Film Institute as one of the most inspirational films of all time, and by the National Film Registry of the Library of Congress as worthy of being preserved for all time for its cultural, aesthetic, and historical significance, this film is about what ordinary people do to reject fear, pull together, and carry on in the face of relentless peril.

At any time during your preparation efforts—perhaps after watching the Red Cross video *A Family Guide to First Aid and Emergency Preparedness*, or maybe when you find yourself or your family members thinking about scary threats—take a break and watch *Mrs. Miniver* to remind yourselves that the most important aspect of preparedness is a state of mind in which we courageously cultivate hope, not fear.

How Much Does It Cost?

Throughout this book you will find descriptions of specific products, with dollar signs indicating about how much they cost. Prices change all the time and may vary from seller to seller, so we have given a range rather than an exact price. We have used the following symbols to indicate price ranges:

$	Below $20
$$	$20–$50
$$$	$51–$200
$$$$	$201–$500
$$$$$	$501+

CHAPTER 1

Preparation Basics

Preparation Basics

Preparing for a disaster requires imagination and logic, in that order. When you think about facing a disaster, what image comes to mind? The brain's go-to image is different for each person, and in great part depends on where you live and the most recent news footage or photos you have seen. But even if you limit your expectations to a regional disaster, say an earthquake if you're in California or a tornado if you're in Oklahoma, you have only begun to imagine what that disaster means to you and yours, because what you do in that moment depends on where you are.

When disaster strikes, you may be at home, at work, or on the road. Your children may be at school. For unprepared disaster victims, being separated from family members, both human and animal, can dramatically increase stress, limiting the ability to make good decisions in the moment. But key to disaster preparation is a family communication plan, enabling loved ones to quickly find one another in an emergency. The plan should include a primary and a secondary meeting place, so that each family member knows where to go after a disaster. It should also include an out-of-state friend or relative designated as a personal call center, since after an emergency it is typically easier to call long distance than locally.

Basic Steps to Be Prepared

1. Imagine your home and workplace in a worst-case scenario, where there are no emergency services, no running water, no power, and no functioning grocery stores.

2. Think about what you will need to evacuate and what you will need if you must stay where you are. You must be ready for either scenario.
3. Create a family communication plan, with a primary and a secondary meeting spot and an out-of-state friend who will relay calls.
4. Discuss and review your communication plan and what each family member will do in the event of an emergency.
5. Assemble a 72-Hour Kit.
6. Pack a Go-Bag for each family member and pet.
7. Learn first aid and become certified (see chapter 5).
8. Get certified to be active in CERT (Community Emergency Response Team).
9. Keep a minimal emergency kit in your car and another one at your workplace (see chapters 13 and 14).
10. Mark your calendar to replace emergency consumables every six months.

Preparation also means having on hand sufficient emergency water rations and the right food—not just any food, but items carefully selected to meet the particular dietary needs of your clan. Imagine going without food or being given food to which someone in your family has a severe allergy. With good planning, you and yours will have water, food, shelter, and warmth after a disaster.

Each individual and family need to prepare both a Go-Bag and a 72-Hour Kit. Your Go-Bag will have what you need if you are being evacuated in advance of a disaster—for example, an oncoming hurricane or possible flood—to a place where you know some basic supplies will be available, such as food and water. Your 72-Hour Kit should contain everything you need to survive on your own after an emergency. This means having enough food, water, and other supplies to last for three days. Relief workers will be on their way after a disaster, but they cannot reach everyone immediately. You could get help in hours or days. Basic services such as electricity, gas, water, sewage, and even cell phone service may be cut off. Some sources suggest you have enough for a two-week period. This book will guide you through your preparations for both cases.

Your Go-Bag

A Go-Bag is a collection of items you may need if you are evacuated. Pack one for each family member in something that is sturdy and easy to carry, such as a backpack or a small suitcase on wheels. Make sure each Go-Bag has an ID tag, and put it in a safe, watertight container where everyone in the household can easily grab it and go. Your basic Go-Bag should include:

- Water, one gallon per person per day
- Nonperishable food, such as energy or granola bars
- Copies of your important documents in a waterproof and portable container (insurance cards, birth certificates, deeds, photo IDs, proof of address, etc.)
- Extra set of car and house keys
- Credit and ATM cards, and about one hundred dollars in cash (in small bills)
- Flashlight
- Tent
- Battery-operated AM/FM radio
- Extra batteries
- Prescription medication. Also include a list of the medications each member of your household takes, why they take them, and their dosages, and a copy of your health insurance.
- List of allergies to medication or food
- Personal hygiene items, such as toothbrushes, toothpaste, toilet paper, and deodorant
- First aid kit
- Contact and meeting place information for everyone in your family
- Small regional map
- Compass
- Special care supplies for children, seniors, or people with disabilities
- Pet care supplies, if you have pets (shelters may have food for humans but not for pets)
- Lightweight rain gear

- Mylar blanket or space blanket
- Change of underwear
- Spare cell phone charger

Create a Five-Minute Go-List Now. It is much easier to make critical choices when you are not under distress. Walk through each room in your house and make a list of one or two irreplaceable treasures that if lost would cause great emotional heartache, for instance, an heirloom ring or a family photo album. You should be able to grab these items if you were evacuating your home at breakneck speed. Start your list with what you can manage alone. If you have help, you can always assign tasks. You can also create a Ten-Minute or Fifteen-Minute Go-List in case you have extra time. Place the Go-List in your Go-Bag, or if possible, place items in your Go-Bag now. Keeping you and your loved ones safe is your most important priority, but if you plan ahead, you might also be able to save a few sentimental material items.

Your 72-Hour Kit

The 72-Hour Kit can be as basic or as elaborate as you would like to make it. Unless you want to stock up on food for months or years rather than days, the essentials for the 72-Hour Kit can fit in a box that you can stash beside your water containers. If you must leave home, take your Go-Bags with you as well.

Your 72-Hour Kit should be customized to meet the needs of you and your household. Here are some suggestions. While all items are described in detail throughout this book, those in boldface are thoroughly explained in subsequent chapters.

- **Water** and **water purification tools** (see chapter 2)
- **Food**, including pet food, diet-specific foods, and, if applicable, baby formula (see chapter 3)

- **First aid kit**, plus extra prescription drugs, extra glasses, and contact lenses and solution (see chapter 5)
- **Important family documents**, including IDs for all family members and copies of insurance, house, and financial papers in the event of an evacuation, especially in disasters that may destroy your home (see chapter 8)
- **Radio**, hand-crank National Oceanic and Atmospheric Administration (NOAA) weather radio with tone alert (see chapter 8)
- **Flashlight** (see chapter 6)
- Wrench or pliers to turn off utilities
- **Duct tape and plastic sheeting**, essential for sheltering in place, and useful in numerous other situations (see chapter 4)
- Cell phone and charger
- **Fire extinguisher** (see chapter 7)
- Signal flares
- Batteries
- Whistle to call for help
- Manual can opener
- Mess kit or plastic dishware and utensils
- Paper goods: toilet paper, towels, dust mask (or cloth), diapers, moist towelettes
- **Personal hygiene products** (see chapter 9)
- Medications and a list of medications for the household
- Plastic bags with ties
- **Matches in waterproof container** (see chapter 7)
- Clothing (for both weather extremes)
- Bedding (for both weather extremes)
- **Tent** (see chapter 4)
- **Compass** (see chapter 8)
- **Shovels** (see chapter 10)
- Mylar or space blanket
- Maps and/or GPS
- Cash and coins, about one hundred dollars in cash (in small bills)

On Alert

- Make sure all the cash you have put away is in small denominations; it's unlikely anyone will be able to make change for you.
- If you store extra medication in your Go-Bag or 72-Hour Kit, be sure to replace it before it expires.
- If you are at home or near home, you may be able to have more supplies with you. If you must evacuate, you may be forced to take less. It's wise to be ready for both contingencies, rather than trying to decide in the stress of the moment what to take and what to leave behind.

CHAPTER 2

Water

CHAPTER 2

Water

Disasters endanger the safety of drinking water. Sewage pipes and water lines are often severed in severe disasters, contaminating public drinking water sources. For example, in 2005, three days after Hurricane Katrina struck Louisiana with 140 mph winds along the coastline and storm surges above twenty-five feet, almost 80 percent of New Orleans was submerged, and the water was heavily polluted by animal carcasses, petroleum products, industrial chemicals, and raw sewage.

Even if the water supply remains clean, people who live in buildings taller than three or four stories may find themselves without water if the electricity goes out, since water is typically pumped to higher floors.

In an emergency, don't expect to turn on your faucet and drink. Instead, be responsible for stocking up your own clean supply of water. Without food, we can survive for weeks, but deprived of drinking water, we can live for only a few days. So water is the first item to prepare for your 72-Hour Kit. It may be wise to be ready for a possible two-week stay at home with no running water.

Three gallons of water per person per day will be enough for drinking needs and basic hygiene such as brushing teeth and washing fruits, vegetables, and dirty dishes. Figure out how much you'll need based on the size of your family. You can buy commercially bottled water or collect and store water from your tap. If you decide to store tap water in recycled containers, avoid containers that held milk or juice, because the proteins and sugars linger after cleaning, inviting bacteria to grow in your stored water. Instead, use strong plastic containers such as soda bottles with airtight, screw-on lids. Never store your drinking water in glass containers, as they may break.

Store your water supply away from direct sunlight in a moderately cool place. Never store water supplies near stored liquids that have an odor, including chemical products such as pesticides or gasoline, because the vapors from these chemicals can penetrate plastic over time and contaminate your water.

Label your stored water with the month and year. Some disaster experts recommend you replace your water supply every six months, when you replace your food supplies, and others recommend you replace water by the "use by" date (if you have purchased it). You can also purchase pouches of emergency water with a five-year shelf life, or simply plan to purify your water supply as needed.

Supplies You'll Need: Water

EYEDROPPER
$ | Amazon.com | 72-Hour Kit

An eyedropper is a small glass tube with a rubberized bulb attached to one end. They are available generically from pharmacies, and are sometimes labeled as "medicine droppers" or "eye and ear medicine droppers." Commonly packaged in a set of two, eyedroppers are inexpensive and provide the only measuring unit—the drop—approved by the Red Cross for use in the water purification method using household chlorine bleach.

CHLORINE BLEACH
$ | Amazon.com | 72-Hour Kit

Available in grocery stores and pharmacies, chlorine bleach is used for water purification, as recommended by the Red Cross. It should contain no additives, scents, or dyes. Purchase bleach that contains, as the sole active ingredient, sodium hypochlorite at a concentration of 5.25 to 6.0 percent.

WATER PURIFICATION TABLETS

$ | Amazon.com | 72-Hour Kit

Water purification tablets are available from a number of manufacturers registered with the Environmental Protection Agency (EPA). Select tablets containing at least 5.25 percent hypochlorite as the only active ingredient.

LIFESTRAW PERSONAL WATER FILTER

$ | Amazon.com | 72-Hour Kit

The LifeStraw is like a water bottle with a straw. You fill the bottle with impure water and draw it up through the straw; on the way, it passes through a small filter. The filter blocks particulate matter (as small as 0.2 microns), bacteria (99.9999 percent), and protozoan parasites (99.9 percent). It is lightweight and compact, and can filter about 260 gallons of water—a small amount at a time. LifeStraw will not filter out viruses, chemical contaminants, or salt.

LIFESAVER BOTTLE 4000 ULTRA FILTRATION WATER BOTTLE

$$$ | Amazon.com | 72-Hour Kit

This personal filtering device uses nonceramic filters with fifteen-nanometer openings—small enough to prevent waterborne pathogens from passing through. Up to 99.999 percent of viruses and bacteria are removed, and pesticides, endocrine-disrupting compounds, medical residues, and heavy metal residues are reduced. The twenty-five-ounce bottle uses replaceable filters that can filter about one thousand gallons of water each. Lifesaver Bottle is approved by the EPA and the European Drinking Water Directives Council.

MSR MIOX PURIFIER PEN

$$$ | Amazon.com | 72-Hour Kit

Miox Purifier works by passing an electrical charge through a small sample of untreated freshwater (lake water, for example) mixed with salt. The electrical charges and brine together create a solution that, when added to untreated water, inactivates viruses, bacteria, giardia, and cryptosporidium. It is ranked as the fastest-acting purification system for large quantities of water, and the only system that provides a means of safety-testing treated water before consumption. Miox was developed for military use and is EPA-approved. It runs on camera batteries and table salt, weighs 3.5 ounces, and is seven inches long.

5-GALLON WHITE BUCKET AND LID (SET OF 3)

$$ | Amazon.com | 72-Hour Kit

These heavy-duty, two-millimeter-thick buckets are made from food-safe plastic and are BPA-free. A set of three in a 72-Hour Kit is a good idea, since they are good not only for retrieving and transporting potable and nonpotable water, but also can be modified for use as an emergency latrine (see chapter 9). These buckets can sometimes be obtained for free from restaurants, and sterilized using a bleach and water solution.

WATERBRICK 3.5-GALLON WATER CONTAINER

$$ | Amazon.com | 72-Hour Kit

WaterBrick containers are a super convenient but more expensive option for water storage. The blue, BPA-free, FDA-approved plastic bricks are stackable and fit together for long-term storage, but can be grabbed and carried using the attached handle. A spigot (purchased separately) can be attached to each brick to fill smaller containers.

WATERBRICK WATER STORAGE CONTAINER SPIGOT

$ | Amazon.com | 72-Hour Kit

Water is precious and never a thing to waste, particularly in the midst of an emergency. This plastic spigot makes it easy to pour out water—no leaks, no spills. Made to fit the WaterBrick, the WaterBrick Spigot screws on perfectly once the original lid is removed (and it tightens easily by hand). Bonus: the ventless spigot can also be used for safe food storage.

Boil Water Orders When tap water is contaminated, a "boil water order" is issued by the agency overseeing public water supplies. When the order is rescinded, it is safe to drink tap water again.

How to Purify Water

After a disaster, water from an unsecure source can be contaminated even when it does not smell, taste, or look contaminated. Unless you are certain the water you are about to drink is pure (such as the clean supply you prepared before the disaster), you must treat it to kill any bacteria or microorganisms that cause disease. Here is the Red Cross postdisaster water treatment protocol.

- Filter solid particles from the water using a coffee filter or a scrap of fabric.
- Boil this filtered water for about one full minute.
- **Let the water cool.** If it is not allowed to cool, the purification will not work!
- Add sixteen drops of liquid chlorine bleach per gallon of water, or eight drops per two-liter bottle of water. Stir to mix. Sodium hypochlorite at a concentration of 5.25 to 6 percent should be the only active ingredient in the bleach, which should contain no fragrance, dye, or soap.
- Let stand thirty minutes.

- If the water smells of chlorine, you can drink it. If it does not smell of chlorine, add the same amount of bleach again and let stand for thirty more minutes. If the water still does not smell of chlorine, discard it and find another source of water.

Stick with Bleach The Red Cross advises using plain bleach to purify drinking water. Do not rely on iodine to purify water, because it will not remove cryptosporidium—protozoans that are very difficult to destroy and that will make you very sick for a very long time. Also, do not use purification products sold by camping or surplus stores, or their websites, unless the products contain at least 5.25 percent hypochlorite as the only active ingredient.

On Alert

- The water in your household water pipes, as well as what's in your water heater storage tank, can be drained and purified for drinking. Turn off the gas or electricity for the heater, drain, and treat the water using the Red Cross postdisaster water treatment protocol.
- For personal hygiene but not for drinking, you may use water from a toilet tank (not the toilet bowl) if no chemicals have been added to the tank water. You can also use hot tub or swimming pool water.
- Do not attempt to purify or drink water from a waterbed, from a home heating system boiler, or from a radiator. All of these water sources are potentially tainted by chemicals that cannot be removed by purification.

CHAPTER 3

Food

CHAPTER 3

Food

Open, well-stocked supermarkets and takeout restaurants are things we take for granted, until we face an emergency. In a disaster, you could find yourself housebound for days; even if you're able to get out and about, the places where you buy food may be closed. Carefully selected reserves of emergency food can provide both nutrients and morale-boosting comfort during a tough time, and ensure that those with special dietary needs do not face extra challenges during a disaster. This includes your beloved animals, of course.

When selecting foods to store, begin by taking inventory of your family members' ages and diets. Babies, the elderly, and those with food allergies must have their special dietary needs met during an emergency, as must dogs, cats, birds, and other companion animals.

In addition, set aside some small celebratory treats to brighten any rough patches during an emergency. However, do practice tough love when it comes to foods and beverages containing sugar and caffeine, since both are dehydrating and will increase water needs during an emergency. And limit very salty foods, which increase thirst.

As for calculating quantities, each family will be unique, and whether you have stored enough food will depend on the circumstances of the disaster. Food rationing during an emergency is fine for everyone except children and pregnant or nursing women. The rest of us, unless we really exert ourselves physically, can function quite well eating half of what we normally eat in a day. But don't forget to make that smaller amount of food as nutrient-dense as possible; even though we can live on less food, it's especially important to keep ourselves as healthy as possible during a disaster.

Food Plan A

Plan A is the universally recommended emergency food ration plan that makes use of items in cans, jars, bottles, boxes, and bags that you can buy at just about any supermarket and set aside for emergencies. It is simple enough to do the first time, and will afterward require only a note on your calendar to rotate your stock throughout the year.

The most important thing is to tailor your food supply to the needs of the people in your family. The most vulnerable populations are the elderly and children. Does anyone in your family have a food allergy? Is anyone diabetic or prediabetic? Make sure you include infant formula and pet food. Whatever food challenges you and your family members face day to day must be addressed as you create your store of emergency foods.

Managing food supplies before and during a disaster does require some thought. These tips from the Red Cross will keep your Plan A food stash at its best.

- Keep food in a dry, cool spot—a dark area, if possible.
- Open any food boxes and other resealable containers carefully so that you can close them tightly after each use.
- Wrap perishable foods, such as cookies and crackers, in plastic bags and keep them in sealed containers.
- Empty open packages of sugar, dried fruits, and nuts into screw-top jars or airtight canisters (with BPA-free plastic) for protection from pests.
- Inspect all food for signs of spoilage before use.
- Throw out canned goods that become swollen, dented, or corroded.
- Use foods before they expire (write the date on the label with a marker), and replace them with fresh supplies.
- Place new items at the back of the storage area and older ones in front.

Smaller Is Better When purchasing canned and packaged foods, aim for smaller items that can be finished quickly after they are opened. Storing leftovers may be difficult if there's a power outage. For more storage guidelines, see Appendix A.

Food Plan B

Everyone's different. If the idea of assembling all kinds of foods and putting together balanced meals in a crisis makes your head spin, or if maintaining an ongoing rotation of emergency food rations seems like way too much work, there are manufacturers who invite you to buy your way out of these tasks. As the prepper population expands, so does the emergency products industry—including the availability of convenience meals alleged to have a twenty- or thirty-year shelf life. But such convenience comes at a price. A year's supply of dehydrated meals for four people, for example, is $4,000, and comes on a pallet of thirty-six plastic six-gallon buckets. See Appendix A for a sample of emergency full-meal products.

Don't Put All Your Dehydrated Food in One Basket: Diversify Storing a variety of food will help to balance out cooking options, keep your menus as fresh as can be, and prevent food burnout. In addition to storing dehydrated and/or freeze-dried foods, try to add store-bought canned goods, or better yet, home-canned goods. Following the same vein, do not place all of your stored food in one location. It is wise to designate a couple of safe locations (dark, cool, and dry spots) in case something happens to one batch; think flood or fire.

Supplies You'll Need: Food and Storage

MAINSTAY EMERGENCY FOOD RATIONS

$ | Amazon.com | Go-Bag | Car Kit

Emergency food bars are great for your Car Kit and Go-Bag. Read the ingredients before purchasing, and make sure the bar contains good, nutritional ingredients and no fillers. During an emergency your body will need a calorie-dense food that is not loaded with sugar or other nonnutritive fillers. This Mainstay pack of nine high-energy lemon-flavored bars (about 400 calories each) has a five-year shelf life and is packaged to withstand temperatures from minus 40 to 300 degrees Fahrenheit without spoiling. The manufacturer says this bar exceeds U.S. Coast Guard standards, is enriched with vitamins and minerals exceeding the RDA requirements, contains no tropical oils or cholesterol, and is kosher and halal.

RUBBERMAID EASY FIND LID 24-PIECE FOOD STORAGE CONTAINER SET

$$ | Amazon.com | 72-Hour Kit

Rubbermaid designed these lids to snap together or onto the base of the container, keeping everything organized. The twenty-four-piece BPA-free set includes two containers of each size: 0.5, 1.25, 2, 3, 5, and 7 cup, with the same lid accommodating bases of several different sizes. For dry packaged foods in your 72-Hour Kit, these containers have tight-fitting lids that will help your food supplies last longer after they have been opened. Because the plastic is see-through, you can quickly find the foods you are looking for and know how much you have on hand. Keeping your food in such containers will also discourage mice and other animals, which are known to smell food and eat through a cardboard cracker box, for example.

LOCK & LOCK STORAGE
$ | Amazon.com | 72-Hour Kit

Storing food in airtight containers extends its shelf life and keeps out pests. Avoid glass containers, which can break. Instead, look for BPA-free plastic containers made of number 2 HDPE (high-density polyethylene), number 4 LDPE (low-density polyethylene), or number 5 PP (polypropylene). BPA-free plastic is reportedly safer for storing food because it reduces the risk of leaching endocrine-disruptive chemicals into wet foods. This 16.75-cup container has a watertight sealing lid. Other sizes are also available.

BOY SCOUTS OF AMERICA ALUMINUM MESS KIT
$ | ScoutStuff.org | 72-Hour Kit

This single-person aluminum mess kit weighs 1.1 pounds and contains one seven-inch frying pan with a steel folding handle, one 0.75-quart pot with a lid, one seven-inch plate, and one eight-ounce plastic mug. It all fits into a mesh carrying bag. You can both cook and eat with all the items except the plastic cup. The pot, pan, and plate are not big enough to make a family meal, but they are just right for one.

LIGHT MY FIRE 6-PIECE OUTDOOR MEAL KIT
$$ | Amazon.com | 72-Hour Kit

You can't cook with this mess kit, but it does have the advantage of a sleek Swedish design that comes in a dozen colors. The Light My Fire mess kit weighs twelve ounces and measures 9 × 8 × 3 inches when it's packed. It contains two plates, a cup with a lid, a colander–cutting board, a spork (combination spoon and fork), and a small waterproof box. The kit is made of BPA-free polypropylene, except for the spork, which is made of polycarbonate. It is watertight, dishwasher safe, and floats. The plastic does not soften when it holds hot foods.

SIMPLE PORTABLE LIGHTWEIGHT SOLAR PANEL COOKER
$$ | SolarOvens.net | 72-Hour Kit

While you may decide to cook on a camping stove or the barbecue when the power goes out, you might want to investigate using a solar cooker, which requires no fuels, emits no fumes, and adds no smoky flavors to foods. Solar cookers cook food more slowly than conventional ovens, but they can be used to pasteurize drinking water in the absence of fuels. This small-capacity cooker weighs eight ounces and can reach about 250 degrees Fahrenheit under normal sunny conditions. It folds flat (14 × 14 × 2 inches). It includes a clear, reusable, high-temperature cooking bag, cooking instructions, and several recipes.

GLOBAL SUN OVEN
$$$$ | Amazon.com | 72-Hour Kit

The Global Sun Oven weighs twenty-one pounds, measures 19 × 19 × 11 inches, and can cook and bake at temperatures up to 400 degrees Fahrenheit using only sunlight. It works in all seasons and has been tested at below-zero temperatures. The reflective surface is rustproof, highly polished, anodized aluminum. A temperature gauge is built into the interior, and there's a carrying handle on the outside.

Do-It-Yourself Solar Cookers Solar cookers are good for the environment and can be very inexpensive to make and to use. SolarCooking.org has plans and instructions for making dozens of different types of solar cookers.

SURVIVAL SEED VAULT
$$ | Amazon.com | 72-Hour Kit

You can plan for months or even years of food self-sufficiency by putting aside seeds for a scenario in which there is time enough to grow your own future meals. Numerous manufacturers of emergency-preparedness products package heirloom seeds (which have not been genetically modified) for emergency kits. This one has twenty varieties of fruits and vegetables, including beans, Swiss chard, watermelon, cantaloupe, broccoli, and cucumbers. The seeds are packaged in triple-layer resealable foil bags placed inside a can. Detailed planting, harvesting, and seed-saving instructions are included.

FOOD STORAGE VEGETABLE GARDEN SEED KIT
$$$ | Amazon.com | 72-Hour Kit

This kit contains 1.4 pounds of seeds—enough for several large gardens—packed in resealable triple-layer foil bags inside a number 10 can. The seeds have been specially dried to give them a long shelf life and a high germination rate. The kit contains sixteen types of nonhybrid vegetable seeds, including corn, onions, spinach, lettuce, beets, carrots, peppers, and zucchini, plus a booklet that explains how to grow your garden.

Camp Cooking If you have to cook without household electricity or gas, you can use your fireplace to cook indoors, but always make sure flames are extinguished if you plan to leave the room for any amount of time. For outdoor cooking in an emergency, use a camp stove, a solar cooker, or a charcoal grill. You may want to keep a small barbecue grill and a supply of dry charcoal with your emergency supplies.

On Alert

- If the electricity fails, first use up the perishable foods from the refrigerator, and after that the frozen foods. When all the perishable food is gone, switch to eating the nonperishables before you turn to your emergency food bars.
- If you store canned foods, make sure you have a manual can opener stored with your food. You can heat canned foods in the can, after first removing the label and opening the can.
- Do not reuse the thin plastic bottles you buy water in, because the plastic breaks down with reuse, leaching chemicals into beverages and foods.
- Be sure to include vitamin, mineral, and protein supplements in your stockpile to ensure adequate nutrition.

CHAPTER 4

Shelter

CHAPTER 4

Shelter

I n the parlance of emergency professionals, in a disaster you either "evac-uate" or "shelter in place." Within those two options, though, there can be countless variables. Circumstances, or a direct order from the authorities, will determine whether your find yourself at home or away from home.

Evacuation means leaving your home and sheltering at the home of some-one you know or at a designated public location. Preparation for evacuation means having a Go-Bag ready, at the minimum (see chapter 1). If evacuation includes driving your car, being prepared will enable you to quickly grab extra items, including Go-Bags, already prepared for each person and animal in your home. If you cannot take your vehicle, you will take only what you can wear and carry. Again, making these decisions in advance is an important part of your preparation so you can get out quickly, with your mind as clear and calm as possible under the circumstances.

It's always stressful to evacuate. In 2013, as raging wildfires burned thou-sands of acres and hundreds of homes in Colorado, those with Go-Bags ready likely experienced less stress during mandatory evacuations—and had with them copies of the paperwork they would need to rebuild their lives if their homes were destroyed by fire.

Sheltering in place means having your 72-Hour Kit (see chapter 1) on hand, including all you will need to keep your family and pets fed and hydrated for at least three days (and up to two weeks), in any temperature at home—or what might be left of your home. If the weather is hot, plan to double the per-person water allotment. If the weather is cold—especially for those living in colder climates—make sure you set aside enough warm clothing and bedding to keep you and your loved ones warm if the heat goes off.

Earthquakes and storms can damage windows, walls, doors, and roofs, so heavy plastic sheeting and duct tape are the universal supplies to have on hand for emergencies that require sheltering in place. Make sure the duct tape you buy for this purpose is of high quality, because cheap brands of duct tape don't stick in wet weather. Likewise, the plastic sheeting you buy should be at least four to six mils thick or thicker, as recommended by the U.S. Department of Homeland Security for sheltering in place during an air-poisoning event such as an explosion at a chemical plant or refinery.

Wherever you live, prepare your home for extreme weather and, yes, earthquakes—even if you don't live in California, since forty-five states and territories of the United States have the potential to experience earthquakes. Learn the facts about natural and other disasters, and how to prepare for them, by visiting Ready.gov.

How Much Is a Mil? A mil is a thousandth of an inch—that is, 0.001 inch. The thickness of many manufactured items is measured in mils. Thirty mils is about the thickness of a credit card.

How to Shelter in Place Outdoors

Depending on the circumstances, you might be able to shelter in a family camping tent, supposing you have evacuated by car or can safely remain near your home but not inside it. If you do not have a tent, you can use space blankets (very thin reflective sheeting, also called emergency blankets or Mylar blankets) or plastic sheeting and duct tape to assemble a basic emergency shelter. You can also buy a Mylar tube tent, but whether you buy one or make your own, the idea is basically the same: a triangular tube is suspended from a nylon cord at the apex. This means you must be able to tie the cord to two upright objects that are far enough apart to accommodate your tent.

Keep in mind that while Mylar will block wind and water, it will also tear quite easily once an edge is cut, so avoid cuts and tears to the edges. As necessary, reinforce the edges of the material with duct tape to prevent tears. Do not discard the used tent, even if there are tears, because the material can be used again for sheltering or for medical emergencies (see chapter 5).

If you don't have Mylar or plastic sheeting, you can attach emergency blankets to a primitive lean-to shelter. A lean-to is a slanted-roof shelter made of whatever you can grab. It can be made of sticks or branches. It leans against a rock or other sticks and branches. It can be made from a tarp, using cording wood tied to a tree. A lean-to breaks the wind coming from one direction, or shades the sunlight if the weather is really hot.

How to Shelter in Place During a Chemical or Biological Emergency

If there is an accident or an attack involving an airborne contaminant, you will shelter in place within a sealed area of your home. Preparing in advance for such an event will protect you and yours more quickly from possible exposure to poisonous chemicals or biological disease agents.

Choose a room in the central part of your house with the fewest number of doors and windows, preferably one with access to running water—for example, a large bathroom. To prepare well ahead of any such event, caulk or duct tape areas around the pipes (if you are using a room with running water) to seal out air. Do not seal off the sheltering room in advance, but prepare the lengths of plastic you will use if there is an air-poisoning event.

Measure all vents, windows, and doors in the room, then cut pieces of heavy plastic sheeting to cover each opening, making each sheet at least six inches wider than the openings you will be covering. Make a covering for every opening, including all heating and air-conditioning vents.

Fold and label each piece of plastic with thick black marker, so in an emergency you can quickly gather all people and pets into the sheltering room and easily duct tape these sheets in place to seal out contaminated air.

 **Stay Informed with a Radio** Keep a battery-operated radio with your precut plastic sheets and duct tape so you can have it with you in your sealed space. This is the only way you'll know when authorities have declared the surrounding air safe again. Rotate the batteries to ensure they're fresh.

Make sure you have turned off any heating or air-conditioning units that would bring in air from outside. Some emergency instructions include placing wet rolled towels along the cracks under doors, but if your room is sealed with plastic, this is not necessary. Do not plan to sleep in your sealed-off room. Officials will inform you by radio when it is safe to exit your temporary shelter. Typically, a contaminated air plume dissipates in less than four hours.

Supplies You'll Need: Sheltering in Place

MARKER, MEASURING TAPE, SCISSORS
$ | Amazon.com | 72-Hour Kit

These common household items are available in most stores and are used to create (before you actually need them) the right sizes of plastic sheeting to store in advance of a potential disaster involving air poisoning. Any standard measuring tape is fine, as are any scissors with which you can cut plastic sheets. If you have windows or doors that exceed the length of a standard tape measure, you can mark where the measuring tape ends and where you start again. Following this logic, you could use a ruler instead. Choose an indelible marker to label your plastic sheets.

HUSKY CLEAR 6-MIL POLYETHYLENE SHEETING
$$$ | HomeDepot.com | 72-Hour Kit

For multiple uses, including sheltering in place during a chemical or bio-
logical attack, the Department of Homeland Security recommends using
clear plastic sheeting of a thickness between four and six mils or thicker. How
much sheeting you need depends on the areas you need to seal. This package
is 20 × 100 feet and comes on a roll.

CLEAR PLASTIC POLY SHEETING, 10 MIL
$$$ | Amazon.com | 72-Hour Kit

Plastic sheeting is taped over windows and doors to keep a room or home safe
during a chemical or biological accident or attack. This roll is 20 × 100 feet
and costs about forty dollars more than the Husky plastic sheeting, but it is
almost twice as thick. The sheeting is wrinkle- and weather-resistant, and has
been used to seal and enclose areas for asbestos abatement.

3M EXTRA HEAVY DUTY DUCT TAPE 6969 SILVER
$ | Amazon.com | 72-Hour Kit

Duct tape is essential in emergencies. When purchased for emergencies,
including sheltering in place during a chemical or biological disaster, the
Department of Homeland Security recommends using duct tape at least ten
mils thick. This is the priciest duct tape, but it is well made and will do the
job. A single roll weighs 1.8 pounds, is 10.7 mils thick, 1.8 inches wide, and
almost 180 feet long. It has a high-tensile strength of thirty pounds and has
been burn-tested. One roll of tape is essential, and two would be better, especially if you have cathedral ceilings, doors, and windows.

EQUINOX POLY TARP

$ | EMS.com | 72-Hour Kit

This general-use, blue, waterproof tarp with aluminum grommets every three feet, including the corners, measures 12 × 14 feet. It's made of high-density woven polyethylene scrim coated with low-density polyethylene. A polypropylene rope reinforces the hem. A poly tarp is an all-purpose tool that can be used primarily to keep rain or wind out of a home where windows have been broken. It can be attached over broken windows with duct tape. It can also be used to make a lean-to.

ROTHCO 550 LB. TYPE III NYLON PARACORD

$ | BigFlySports.com | 72-Hour Kit

Strong nylon cording will have multiple uses in your 72-Hour Kit, not the least being to tie down a tarp for shelter or shade. It also makes a good clothesline and can keep a gate closed if necessary. This cord is one hundred feet long and comes in sixteen colors and patterns. It is $\frac{5}{32}$ inches in diameter and has a seven-strand core. It's made by a certified U.S. government contractor and has a 550-pound test strength.

EMERGENCY ZONE BRAND EMERGENCY SHELTER TENT

$ | Amazon.com | Go-Bag | 72-Hour Kit

If you can't shelter inside your home or find a public shelter, you may use a family camping tent, or create shelter from plastic sheeting or Mylar blankets and duct tape. Or you may decide to keep a small, inexpensive emergency tube tent in your Go-Bag or 72-Hour Kit. A tube tent is a triangular tube. This lightweight, two-person tube tent is basically a Mylar tube taped into a triangle. The reflective material helps conserve body heat. It is eight feet long, weighs 7.7 ounces, is waterproof, and comes with an attached nylon cord to set it up. It comes sealed in a waterproof pouch.

ADVENTURE MEDICAL KITS SOL ESCAPE BIVVY
$$ | Amazon.com | 72-Hour Kit

This is basically a waterproof sleeping bag and one-person shelter in one. The bag is 36 × 84 inches, with waterproof seams plus a drawstring hood closure and side zip, so you can seal out the elements entirely or use it as a traditional sleeping bag. The inner fabric reflects body heat inward but lets moisture escape, so condensation doesn't build up inside the bag. The outside is bright orange, making you easy to spot. The bag weighs just 8.5 ounces.

Duct Tape Fashion Duct tape has become a fashion accessory and is now available in many colors and patterns to decorate everything from handbags and wallets to clothing. Stay away from fashion duct tape and stick with the heavy-duty kind. It will adhere and hold up better in emergencies.

On Alert

- To avoid postdisaster fires and explosions from natural gas leaks, make sure everyone in your household knows how to shut off the gas to the house. Consult your gas company and make sure you have the right tools for your situation. Practice with your family, but don't actually turn off the gas—if you do, *do not* attempt to turn it back on. If you smell gas or hear a blowing or hissing noise, open a window and get everyone out quickly. Turn off the gas, using the outside main valve, if you can, and call the gas company from a neighbor's home.
- Family members should learn where and how to shut off the water and electricity, so sparks cannot ignite a gas leak and no one gets electrocuted. When shutting off electricity, turn off individual circuits before shutting off the main circuit. And when shutting off the water, close the main house valve (not the street valve) to avoid the gravity effects

that might otherwise drain the water from your hot water tank and
toilet tanks.

- Never use a candle or conventional flashlight when checking for gas leaks
  or attempting to turn off the gas after a disaster. Protect yourself from
  sparks that, when in contact with gas fumes, can cause a deadly explosion
  and fire. For checking gas valves and potential leaks, buy an intrinsically
  safe flashlight designed for hazardous circumstances (see chapter 6 for
  information about flashlights).

CHAPTER 5

First Aid

CHAPTER 5

First Aid

hen an earthquake measuring 5.0 on the Richter scale hit Napa Valley, California, in 2000, the intensity and duration of the shaking, as well as the severity of the damage, was relatively minor compared with the 1989 Loma Prieta earthquake. But it was a jolting quake nevertheless: structures buckled, people were injured, and gas lines were ruptured. Local 911 calls came flooding in, and almost immediately firefighter response was running two hundred calls behind. Local services were not sufficient to meet the community's needs, and outside help had to be called in from other counties.

During emergencies, rescuers make lifesaving their first priority; threats to the environment and to property come second. The calls for help that come from institutions such as schools and hospitals or care homes for the elderly, where the highest concentration of lives are at risk, take precedence over calls from individual residences. This makes it critical that you and other adults in your household have first aid training, at the least.

Also, the emergency medical needs of any community are triaged: the most life-threatening cases are seen first, and even then, only when and where treatment from a professional is an option. Doctors, nurses, and EMT specialists themselves may be hurt or unable to travel, or they may be called to assist at institutions, or, if there is a disaster-management team at work, wherever they are needed most. Consequently, in a disaster there may be situations where it's truly up to you to save a life. In such cases, families will be responsible for performing first aid when the need arises. Remember, a disaster demands much of us. This is why we prepare.

It takes more than even the most well-stocked first aid kit to help some-one in trouble. To be prepared for any disaster, you must know how, why, and when to use everything in the kit, as well as techniques that don't require a kit—from CPR (cardiopulmonary resuscitation) to psychological first aid for traumatized adults and children. You can learn all this from a first aid course and from CERT training.

In chapter 1 you learned that the first step in preparing for a disaster is to imagine the unimaginable. The same person who is squeamish at the sight of blood under normal circumstances might, in a disaster, be the only per-son who can sew up a wound to save a life. You might temporarily lose your squeamishness if you can save another person, especially someone you love. That's why getting first aid training is so important.

In basic first aid classes, taught by the Red Cross or local fire departments and other organizations, you practice on plastic dummies so that your first attempt at CPR is not in the life-or-death circumstances of an emergency. You may or may not make it to advanced training, but with some founda-tional first aid training, you will at least know you can follow the advice ascribed to the legendary Greek physician Hippocrates: "First, do no harm."

Supplies You'll Need: First Aid

WHERE THERE IS NO DOCTOR: A VILLAGE HEALTH CARE HANDBOOK
$ | Amazon.com | 72-Hour Kit

In extreme circumstances around the world, untrained people have had to perform emergency medical care, including surgical procedures. The globally useful book *Where There Is No Doctor* was written by David Werner, Jane Maxwell, and Carol Thuman and published by the Hesperian Foundation (revised edition). Originally written for inhabitants of rural villages where, as the title suggests, there is no doctor, this unique medical book gives instruc-tions for procedures that normally only a doctor would attempt. However, in disaster situations, anyone might be called on to perform a complex medical procedure to save a life. Even a community that was once a city can be turned into a collection of "villages" cut off from one another and from standard emergency services.

THE AMERICAN RED CROSS FIRST AID AND SAFETY HANDBOOK
$ | Amazon.com | 72-Hour Kit

This book by Kathleen A. Handal is published by the American Red Cross and covers the categories used in standard Red Cross first aid training modules. By following the instructions here, a person could diagnose and treat, in some fashion, the symptoms of someone suffering from anything from poisoning to heart attack to spinal cord injury, stroke, or seizures. Included are sections for making the household a safer place and for surviving disasters.

EMERGENCY MYLAR THERMAL BLANKETS (PACK OF 10)
$$ | Amazon.com | Go-Bag | 72-Hour Kit

Experts strongly recommend having Mylar blankets in emergencies. Getting more than one is essential because they are not built for endurance. Also called "space blankets," these Mylar sheets are great for reflecting solar radiation and helping to prevent hypothermia. But if you nick the edges, they will tear quite easily. If you want to use a Mylar blanket as a makeshift tent and you make a hole in it, the hole will run out to the edge. Use duct tape when placing Mylar blankets under tension.

A FAMILY GUIDE TO FIRST AID AND EMERGENCY PREPAREDNESS
$ | RedCrossStore.org | 72-Hour Kit

To help everyone feel a part of the preparations, make some popcorn and call everyone in for family video night. This Red Cross DVD was designed to help families prepare for everyday medical emergencies, including those caused by disasters. The DVD takes a positive approach that involves training children, and includes step-by-step demonstrations of first aid techniques. Take it seriously and involve your children—a life may rest in the hands of your kids during a disaster.

DELUXE FAMILY FIRST AID KIT

$$ | RedCrossStore.org | Go-Bag | 72-Hour Kit

In a pinch, this bag provides a starting point, but don't buy it as your only set of supplies for a real disaster. This 12.5 × 9 × 2–inch kit weighs two pounds, and although it contains most of what you will need for your 72-Hour Kit, it does not come with all of the recommended items listed later in this chapter. To prepare for a serious emergency, you would need to bulk up this kit, using the list provided later in this chapter.

BE RED CROSS READY FIRST AID KIT

$ | RedCrossStore.org | Go-Bag

When you are in your car or when walking out in a no-car evacuation, this little kit might be what you have in your Go-Bag. Not recommended for sheltering in place, this minimal kit provides supplies for only minor first aid emergencies, and it can be tucked easily into your backpack or auto glove box. It weighs less than a pound and contains mostly bandages and ointment, plus a standard emergency first aid guide.

CAT FIRST AID

$ | RedCrossStore.org | Go-Bag | 72-Hour Kit

If your kitty is feeling poorly, you will want to make sure you can address the illness or injury on your own, since veterinary care will not likely be available. This book was produced by the American Red Cross, and comes with an instructional twenty-eight-minute DVD. It includes giving your cat medication, how to respond to a breathing or heart emergency, and a first-aid reference guide that covers sixty feline medical emergencies. After viewing the demonstrations, you might feel more confident about being a stand-in for the vet.

DOG FIRST AID
$ | RedCrossStore.org | Go-Bag | 72-Hour Kit

Your dog can become injured or ill during a disaster, when you can't take Fido to the veterinarian. Because the onus may be on you to administer animal medicine, this book and twenty-eight-minute DVD produced by the American Red Cross will show you what you need to do to help alleviate your dog's suffering when you can't get to the vet's office. It includes giving your dog medication, how to respond to a breathing or heart emergency, and a first aid reference guide that covers almost seventy canine medical emergencies.

How to Create Your Emergency First Aid Kit

To cover many medical needs during and after a disaster, the following list combines the first aid supply recommendations of the Red Cross, FEMA, and the Mayo Clinic. The complete kit will have enough supplies for four people— although that is a rough estimate, since you don't know how long you will need to rely on the supplies or how badly someone might be hurt. For those who live in remote areas and for anyone who wants to be prepared for long periods without medical services, include a paperback copy of *Where There Is No Doctor* along with additional first aid supplies.

Commercially Available First Aid Kits Many businesses want to sell you a first aid kit, and there are many to choose from with a plethora of claims made about the superiority of the contents. But the best first aid kit is one you spend time personalizing to suit the needs of your family. Simply buying a kit is better than nothing, but it will not prepare your family sufficiently unless all your personal medicines are in it, along with the extra supplies that will at least bring you greater peace of mind during and after a disaster. Do your own research. We have recommended two first aid kits sold by the American Red Cross, but only with the caveat that you customize any kit you buy!

Inside the first aid kit, keep a pencil and a small notebook, so each time an item is used it can be listed and later replaced. Rotate medicines, such as pain relievers, and use them before the expiration dates. You can note this on your calendar and update first aid supplies when you rotate stored perishable foods and water.

Keep your first aid supplies in a bright-red bag made of sturdy waterproof fabric, so it can easily be found when the need arises. Inside the bag, keep a list of family medical issues and a list of the prescription drugs everyone takes. Make sure every family member knows where the first aid kit is stored, and that every person can reach it easily.

Your first aid kit should include the following:

BANDAGES AND TAPE

- 1 roller bandage (3 inches wide)
- 1 roller bandage (4 inches wide)
- 2 absorbent compress dressings (5 × 9 inches)
- 2 triangular bandages
- 5 sterile gauze pads (3 × 3 inches)
- 5 sterile gauze pads (4 × 4 inches)
- 25 adhesive bandages (assorted sizes)
- Adhesive cloth tape (10 yards × 1 inch)
- Duct tape

TOPICAL TREATMENTS

- 2 hydrocortisone ointment packets (approximately 1 gram each)
- 5 antibiotic ointment packets (approximately 1 gram each)
- 5 antiseptic wipe packets
- Petroleum jelly or lubricant
- Sterile eyewash, such as saline solution
- Sunscreen

TREATMENT AIDS

- 1 breathing barrier (with one-way valve, to perform mouth-to-mouth resuscitation)
- 1 instant cold compress
- 1 Mylar blanket (space blanket)
- 2 pairs of nonlatex gloves (size large)
- Cotton balls and cotton-tipped swabs
- First aid instruction book
- N95 respirator masks or surgical masks (1 package)
- Hand sanitzer
- Oral thermometer (not mercury, not glass; test regularly to make sure the batteries work)
- Plastic bags for disposing of contaminated materials
- Safety pins in assorted sizes
- Scissors
- Syringe, medicine cup, or spoon
- Turkey baster or other bulb suction device for flushing out wounds
- Tweezers

OVER-THE-COUNTER DRUGS

- 2 packets of aspirin (81 milligrams each)
- Activated charcoal (use only if instructed by your poison control center)
- Aloe vera gel
- Antacid

- Antidiarrheal medication
- Calamine lotion
- Hydrocortisone cream
- Laxative
- Nonaspirin pain reliever (never give aspirin to children)
- Oral antihistamine, such as diphenhydramine (Benadryl, Sominex, etc.)

PRESCRIPTION DRUGS

- If prescribed by your doctor, drugs to treat a severe allergic reaction, such as an autoinjector of epinephrine (EpiPen, Twinject, etc.)
- Prescription medications you take every day, such as insulin, heart medicine, or asthma inhalers (anything that does not require refrigeration; periodically rotate medicines to account for expiration dates)
- Prescribed medical supplies, such as glucose and blood pressure–monitoring equipment and supplies

On Alert

- To find your nearest CERT training site, visit CitizenCorps.gov. Online training is available, but classroom instruction is an integral part of the certification process.
- Learn and teach others. Take a hands-on first aid course and maintain your first aid certification. And don't keep the knowledge to yourself. In an emergency, your life may depend on the quick response of a spouse or even a child. Teach family members to identify the signs of distress—from choking, heart attack, and stroke to dehydration and other conditions—and what actions to take, even if it is a child alerting an adult to act.
- To educate family members, consider viewing one of the Red Cross educational aides, such as the DVD *A Family Guide to First Aid and Emergency Preparedness*. Make it a practice to watch it at least once a year, so everyone can review what they should do in an emergency.

CHAPTER 6

Light and Power

CHAPTER 6

███████

Light and Power

The biggest blackout in U.S. history occurred in late summer of 2003, leaving forty-five million people in eight states without power for days. In 2012, the largest Atlantic storm in recorded history, Hurricane Sandy, demolished thousands of homes along the eastern seaboard. It affected twenty-four states, claimed 285 lives, and left millions without power for weeks in wet, cold weather. More than 4,100 utility workers from twenty-one states had to be called in to restore power to New Jersey.

While the power was out, life was not easy. A few well-prepared families had the advantage of expensive stationary generators that switch on automatically when the power fails. Others generated electricity with gas-powered (and ear-splitting) mobile generators. Others resorted to fireplaces for heat and for primitive cooking indoors, while some used camp stoves and barbecue pits outside, though it was hardly the kind of weather for comfortable grilling. (For more on cooking, see chapter 7.)

To prepare, take a look at how you use electricity in your home and the combined wattage use. What can you live without when forced to use a generator? By minimizing your consumption during a disaster, you can satisfy your power needs modestly until utility power has been restored to your home.

When shopping for an emergency power generator, consider the wattage needs of your home, the amount of generator fuel you can safely store for an emergency, the amount of generator noise that can be tolerated by family and neighbors, and whether the generator needs to be linked to an automatic switch that starts it the moment the power fails—even if you are not at home.

The switch alone costs about $500, professionally installed, which must be factored into your decision.

· When you're assembling your 72-Hour Kit, don't just throw in an ordinary flashlight; it can be dangerous in hazardous conditions. When purchasing an emergency flashlight, the key phrase to look for is "intrinsically safe" (see below). Intrinsically safe flashlights come in a range of prices, but the most basic one will do fine.

In your 72-Hour Kit, you will also want to include a deck of cards and a few board games or read-aloud books to keep everyone entertained while you wait for power. Children and game-loving adults will discover that playing electronic games will drain batteries relatively quickly, and time without light and power can pass very slowly.

Battery Supply Management As you assemble the emergency devices for your 72-Hour Kit and Go-Bags, keep a complete list of all the batteries you will need and in what sizes. Then calculate how many extras you should have on hand. When purchasing batteries, check the manufacture date, and note it on your list of emergency supplies to rotate.

Intrinsically Safe Flashlight

When a storm or an earthquake leaves you in the dark at home, never reach first for a box of matches, a candle, or a traditional flashlight. Pick up your special emergency flashlight—an intrinsically safe model. This means the light has been designed for use in hazardous situations, including gas leaks. An intrinsically safe flashlight will not create sparks when it is switched on; even the tiniest spark can set off an explosion or fire when used around explosive substances, such as leaking or broken gas lines in your home.

With your intrinsically safe flashlight, check the household gas lines to make sure there are no leaks. If you suspect more earthquake activity or a posthurricane tornado, you can then (if you have the skills to do so) safely turn off the gas to your home. When you are certain there are no gas leaks, you may then turn to other emergency light sources and address other needs.

Supplies You'll Need: Light and Power

ENERGIZER 1-LED HANDHELD INTRINSICALLY SAFE FLASHLIGHT
$ | Amazon.com | Go-Bag | 72-Hour Kit

This inexpensive, intrinsically safe flashlight is a good choice for a basic emergency. The single white LED light provides plenty of light. The body is waterproof and oil and grease resistant, and it has a built-in carabiner belt clip and lanyard. It can survive a three-foot drop and has a shatterproof lens. This flashlight has been approved by the Mine Safety and Health Administration (MSHA). The two required AA batteries are not included.

ENERGIZER INTRINSICALLY SAFE 3-LED HEADLIGHT
$$$ | Amazon.com | Go-Bag | 72-Hour Kit

This three-LED lamp is attached to an adjustable headband for hands-free light. The headlight is waterproof, shatterproof, and survives a six-foot drop test. It has a high and low white beam, plus red for night vision, green for inspection work, and a flip-up lens for diffuse, spot, or area light. The light pivots to direct the beam where you need it. The light is in the front of the headband and the battery pack is in the back to distribute the weight evenly. It runs twenty hours on high using three AA batteries, which are included.

STREAMLIGHT 76411 POLYSTINGER LED HAZ-LO INTRINSICALLY SAFE RECHARGEABLE FLASHLIGHT

$$$ | Amazon.com | Go-Bag | 72-Hour Kit

Developed for hazardous conditions, this intrinsically safe rechargeable LED flashlight runs fifty thousand hours, or twenty days continuously in "moonlight" mode. The lens is unbreakable polycarbonate with a scratch-resistant coating. Its deep-dish parabolic reflector produces a tight beam with good peripheral illumination. The head-mounted on-off switch is designed for an extremely long life. This light has been ten-foot impact-resistance tested and is also water-resistant. It comes with a 12-volt DC charger.

CYALUME SNAPLIGHT INDUSTRIAL GRADE CHEMICAL LIGHT STICKS

$ | Amazon.com | Go-Bag | 72-Hour Kit

Light sticks—or glow sticks—are the safest light to use immediately following a disaster because they do not spark and they are entirely waterproof. Glow sticks are also great to have on hand in case you run out of batteries, because they require no power source (they're powered by a chemical reaction you initiate when you open them). The only drawback is that they are not as bright as a flashlight. These green six-inch tube lights have a shelf life of five years from the manufacture date, and they last twelve hours once lighted. They come in a pack of ten.

RAYOVAC SPORTSMAN LED LANTERN (SE3DLN)

$$ | Amazon.com | Go-Bag | 72-Hour Kit

This little lantern weighs just 14.4 ounces and runs on three D batteries, which are not included. It produces light from a 4-watt LED bulb. It will run forty hours continuously on high mode, ninety hours on low, and has a strobe mode function. You can set it so a green LED blinks every five seconds when the lantern is off, making it easy to find in the dark. The lantern is water-resistant. However, this is not an intrinsically safe light source.

RANGE KLEEN WKT4162 82-BATTERY ORGANIZER WITH REMOVABLE TESTER
$ | Amazon.com | 72-Hour Kit

This 12 × 12 × 1.8–inch plastic organizer can help you locate and test your backup batteries in an emergency. It holds eighty-two batteries of assorted sizes, including AA, AAA, C, D, and 9-volt batteries, plus hearing aid and watch-size batteries. It mounts on a wall or fits in a drawer. You supply the batteries—this is just a case with a removable battery tester.

UCO ORIGINAL COLLAPSIBLE CANDLE LANTERN
$ | Amazon.com | Go-Bag | 72-Hour Kit

This collapsible candle lantern folds down to smaller than a soda can, but pops open to accommodate a nine-hour candle. The sliding glass chimney creates a windproof environment that is also much safer than an open flame if the candle is dropped. A spring-powered candle tube pushes the candle up as it burns, keeping the flame at a constant height.

UCO 9-HOUR CANDLES (20 PACK)
$$ | Amazon.com | Go-Bag | 72-Hour Kit

This is a twenty-pack of candles for the UCO candle lantern. Each candle burns for about nine hours and throws off both heat and light. The wax is designed for minimal dripping. UCO also makes three-packs and packs of citronella candles. While burning these candles in a candle lantern is safer than an open flame, lit candles are still a fire hazard and should never be left unattended.

100-HOUR PLUS EMERGENCY CANDLE CLEAR MIST

$ | Amazon.com | Go-Bag | 72-Hour Kit

This candle is technically a small lantern with a sealed fuel source that supports one hundred hours of odorless and smokeless candlelight. The flame can be blown out and relighted, as a candlewick would. Remember that all fire sources, such as candles and artificial candles, are a fire hazard and should be used only when conditions are safe.

KOHLER 14,000-WATT 200-AMP WHOLE HOUSE AUTOMATIC TRANSFER SWITCH STANDBY GENERATOR

$$$$$ | HomeDepot.com

This standby generator can power your whole house, and is recommended by *Consumer Reports*. Powered by either natural gas or propane, it automatically turns off and on as needed, shuts down when engine oil is low, and reportedly runs very quietly. The housing is corrosion-proof. Professional installation can be expensive, and the unit requires a fifty-dollar battery, which is not included.

TROY-BILT XP 7000 RUNNING WATTS PORTABLE GENERATOR

$$$$$ | Lowes.com

This portable generator is highly recommended by *Consumer Reports*. The 7,000-watt, gasoline-powered generator has an electric starter, fuel shutoff (which prevents leaks and keeps fuel from getting trapped in the fuel system and spoiling during storage), and shuts down when engine oil is low. It has a power meter, a nine-gallon tank for an average fifteen hours of run time, and a fuel gauge. It comes with the battery for the electric-start feature and includes a starter bottle of engine oil. This machine runs loud; *Consumer Reports* recommends using hearing protection closer than twenty-three feet. Like all generators, this is not safe to operate inside your home.

Bonus Items

3-IN-1 TRAVEL MAGNETIC CHESS, CHECKERS, AND BACKGAMMON SET
$ | YMImports.com | 72-Hour Kit

If you find yourself sheltering without television or the power to recharge electronic devices with games on them, you'll have to do something. This game set needs no power and no batteries. The three-in-one combo has a magnetic board that holds pieces in place while you're playing, and folds up and snaps closed to keep all the playing pieces inside. The chess-checkers board is on one side, the backgammon board on the other. All playing pieces and dice are included.

SCRABBLE FOLIO EDITION
$$ | Target.com | 72-Hour Kit

Scrabble is one of those multigenerational favorite games. This Scrabble set comes in a soft-sided zipper case with a full board, plastic letter tiles, and tile holders inside. The letter tiles snap into place on the folding game board, so you can pack up a game and save it to finish later. A score pad and pencil are included.

Go Solar It is easy to forget about the necessity for outdoor lighting when the power goes out. When the street lamps fade to black, the outdoors can appear daunting. Whether you need to get to your car or access outdoor storage, solar garden stakes will be a lifesaver. As a bonus, they can be found for relatively cheap and can be used year-round, disaster or not.

On Alert

- Never operate a power generator inside your home. The fumes can be deadly.
- Glow sticks can be activated during power outages and attached to animal collars, so dogs and cats can be easily located in the dark.
- Traditional flashlight bulbs have limited life spans. Light-emitting diode (LED) flashlights are more durable and last up to ten times longer than traditional bulbs.

CHAPTER 7

Fire and Fuel

Fire and Fuel

I n a disaster, fire can be an important tool or a terrifying threat. It can provide warming light, cooked meals, and boiled-clean drinking water—be it in the home fireplace or outdoors. A cozy fire can boost morale and even signal for help.

But earthquakes and explosions, crashes, bombs, and lightning strikes can all cause unwanted fires. During Hurricane Sandy in 2012, a huge fire caused by seawater flooding electric lines was whipped out of control by the wind and burned more than one hundred homes, devastating the community of Breezy Point in New York City.

These are the kinds of fires we can't prevent. However, in the United States, kitchen fires cause most home fires and are responsible for 81 percent of all fire-related fatalities. The one amazing and well-documented truth about these unwanted fires is that they are entirely avoidable. So before we run off to buy flint or gather twigs, we first need to know how to stop a non-cozy fire—the kind of fire that takes lives.

Fire as a Threat

Firefighters say the most important lifesaving device you can have in your home is a properly working smoke detector. While some people get confused and irritated by battery-life management and the shelf life of the smoke alarm itself—plus the panic and irritation that results from an alarm that sounds when someone burns dinner—recent advances in technology have produced a new kind of smoke detector with a ten-year life span that comes with a sealed lithium battery reported to also last ten years. The new devices also come with a silencing feature for when the cook burns something and the house gets smoky. The design is meant to save you the work of swapping out batteries or remembering how many years the alarm works; all that's left to do is test the alarm monthly.

In an emergency, when lighting and cooking may be makeshift or improvised, we need to be especially mindful and remember that almost all fires are preventable. Preventing fires in the home means not leaving a fire unattended—ever. Keeping an eye on the fire and a lid on the pot can be a lifesaving habit.

A fire extinguisher is an important part of your 72-Hour Kit, but you will need to know how they work and what kind you need before you order an extinguisher. UL ratings and state fire marshal approvals are listed on the extinguisher, as is what class of fire it can be used on (see *How to Extinguish a Fire* for a list and explanation of the fire classes). Some extinguishers have a ratings number; the ratings number increases with the volume of fire-extinguishing agent contained in the fire extinguisher. So, for example, Class A fire extinguisher ratings run from 1-A to 40-A, and Class B from 1-B to 640-B. The higher the number, the greater the volume of fire-extinguishing agent.

Other extinguishers have state fire marshal approvals but no numerical ratings. These extinguishers are marked C for use on electrical fires, D for use on combustible metal fires (with each extinguisher matched to a type of metal), and K for use in kitchens to fight fires fueled by fat and oils.

The advice of *Consumer Reports* is not to buy aerosol flame retardants, because all the products they tested were unreliable. Instead, buy an approved fire extinguisher manufactured within the last year with a pressure indicator reading "full." Problematically, not all manufacturers have a date of manufacture—this omission is especially unfortunate because chemical extinguishers have a twelve-year shelf life, according to the National Fire Protection Association (NFPA).

How to Extinguish a Fire

FIRE TYPE	EXTINGUISHING AGENT	EXTINGUISHING METHOD
Ordinary solid materials	Water Foam Dry chemical	Removes heat Removes air and heat Breaks chain reaction
Flammable liquids	Foam CO_2 Foam dry chemical	Removes air Breaks chain reaction
Electrical equipment	CO_2 Dry chemical	Removes air Breaks chain reaction
Combustible metals	Special agents	Usually removes air
Kitchen oils	Chemical	Usually removes air

CLASSES OF FIRE

To aid in extinguishing fires, fires are categorized into classes based on the type of fuel that is burning:

- **Class A Fires:** Ordinary combustibles such as paper, cloth, wood, rubber, and many plastics.

- **Class B Fires:** Flammable liquids (e.g., oils, gasoline) and combustible liquids (e.g., charcoal lighter fluid, kerosene). These fuels burn only at the surface because oxygen cannot penetrate the depth of the fluid. Only the vapor burns when ignited.

- **Class C Fires:** Energized electrical equipment (e.g., wiring, motors). When the electricity is turned off, the fire becomes a Class A fire.

- **Class D Fires:** Combustible metals (e.g., aluminum, magnesium, titanium).

- **Class K Fires:** Cooking oils (e.g., vegetable oils, animal oils, fats).

It is extremely important to identify the type of fuel feeding the fire in order to select the correct method and agent for extinguishing the fire.

What Is a UL Rating? UL (Underwriters Laboratories) is an independent, not-for-profit safety consulting and certification company that is approved by the Occupational Safety and Health Administration (OSHA) to perform safety testing. It provides safety-related certification, validation, testing, inspection, auditing, advising, and training services. A UL rating means the product has been independently safety tested.

Supplies You'll Need: Fire as a Threat

FIRST ALERT SA305CN SMOKE ALARM
$$ | Amazon.com

Get a smoke alarm and test it monthly. This one is among the new smoke alarms that uses ten-year lithium batteries (9 volts, sealed). This model comes with a silence button for when the cooking gets a little smoky, and a test button to check it every month. It's tamper-resistant and warns when the battery is running low. The 85-decibel alarm is likely to wake even sound sleepers.

KIDDE KN-COSM-B BATTERY-OPERATED COMBINATION CARBON MONOXIDE AND SMOKE ALARM WITH TALKING ALARM
$$$ | Amazon.com

Not all states require a carbon monoxide detector, but this silent killer is a real danger, and it's a good idea to have a detector. This combination smoke and CO detector has a talking alarm to tell you exactly what type of hazard it has detected. A test button resets the alarm, and a "hush" button lets you easily cancel the alarm. This model does not come with long-life batteries, so you will need to routinely change the three AAA batteries; the unit will tell you when the battery is low.

KIDDE FA110 MULTIPURPOSE FIRE EXTINGUISHER 1A10BC
$$ | Amazon.com | 72-Hour Kit

This fire extinguisher is UL rated, Coast Guard approved, 1-A and 10-B:C, which means it is designed to fight basic fires common to the home, including those involving fabrics, plastics, wood, flammable liquids, and electrical equipment. It is fitted with a pressure gauge that tells you the fire extinguisher is charged. There's also a large, clear instruction label pasted on the side.

KIDDE FX10K KITCHEN FIRE EXTINGUISHER, 82CI
$$$ | Amazon.com | 72-Hour Kit

This fire extinguisher is UL rated. It uses a nontoxic fire-suppressing agent specified for kitchen fires and energized electronic equipment. There is a pressure gauge on top, an easy-to-pull safety pin, and a nylon handle that won't rust. This comes with a wall hanger and is made for a single use.

KIDDE KL-2S TWO-STORY FIRE ESCAPE LADDER WITH ANTI-SLIP RUNGS
$$$ | Amazon.com

This thirteen-foot basic fire escape ladder is a foot wide and can hold up to one thousand pounds when the load is distributed on more than one rung, or 750 pounds on one rung. The rungs are antislip zinc-plated steel, and the rails are nylon strapping. The ladder is flame resistant. It folds up accordion-style, with curved arms at the top that secure it under a windowsill. It also comes in a twenty-five-foot length. It's a good idea to remove the ladder from the box before storing it for an emergency!

On Alert

- Preserving life (not stuff) is your first priority, always!
- Use a fire extinguisher that is appropriate for the type of fire that has started.
- When fleeing a fire, close doors as you exit to slow the spread of fire.

Fire as a Tool

For fire needed in a power outage, stock fire starters and dry fuel in your 72-Hour Kit. Make sure you have the fuel and the waterproof matches (or other fire-starting device, of which there are many) to build a fire even in wet weather. If you are not experienced in building a fire, watch a video or ask a camping enthusiast to show you how. Short of that, buy composite logs (see below), which are made of compressed sawdust and wax or other materials that catch fire relatively quickly. An emergency is no time to learn how to build a traditional wood fire, which can be very difficult at times even for experienced campers and hikers.

If you are sheltering at home and can use your fireplace, make sure the flue is open and that you start your fire at the far back of the fireplace. This helps improve the draw of smoke upward and out of the room.

Whether for cooking, driving, heating, or to use in your generator, a backup supply of fuel can come in handy. According to the American Petroleum Institute, gasoline can be safely stored in approved containers (see the "Storing Gasoline" box below) of less than five gallons each for up to two years. They recommend storing a total of twenty-five gallons or less, and it should be stored at room temperature, away from any sources of heat or sparks, and not in your home or garage. Diesel fuel begins to oxidize as soon as it leaves the refinery, so it lasts for only six to twelve months.

Kerosene does not evaporate as easily as gasoline, making it less volatile. Kerosene has a shelf life of about three months in a plastic container (use one approved for storing gasoline). After about three months, bacteria and mold may begin to form in the container. Sunlight degrades kerosene, so keep it in a dark place.

If you have a propane grill or use propane in your kitchen, the good news is that propane is one of the easiest fuels to store. Propane does not degrade over time, the way gasoline and kerosene can, so it's only the container you need to worry about. Keep it dry and rust free.

A supply of seasoned wood is also a necessity if your emergency plans include using a fireplace, wood-burning stove, or cooking over a campfire. For campfire cooking, charcoal is also a good option. Store it in a watertight container. For more information about storing fuels, visit TheReadyStore.com.

Storing Gasoline Plastic milk jugs, antifreeze containers, and glass bottles are not suitable for carrying or storing gasoline. Some plastics become brittle as they age, and some are chemically incompatible with gasoline. Gasoline expands and contracts as its temperature changes, and some containers cannot withstand the pressure of that change in volume. Some containers that are actually sold as gas cans cannot be sealed well enough to prevent spilling.

The best containers for gasoline are safety cans approved by UL or Factory Mutual (FM). Safety cans are available in several sizes and have various mechanisms for opening the valve to pour the gasoline. Some have a funnel spout to make pouring easier and reduce spills. Although the better ones cost more, they are much safer and will last longer. You will find a gas can recommendation in chapter 13.

Supplies You'll Need: Fire as a Tool

UCO STORMPROOF MATCH KIT WITH WATERPROOF CASE

$ | Amazon.com | 72-Hour Kit

Household matches can get soggy, and wet matches will not light. For your 72-Hour Kit, consider this set of twenty-five "stormproof" matches. They come in a waterproof container that floats, and it includes three striking surfaces that slide into a holder on the side of the container. Each match is reported to be windproof and will burn about fifteen seconds before going out, even underwater. Refill packs of matches are also available.

LIGHTNING NUGGETS N100SEB FIRESTARTERS

$$ | Amazon.com | 72-Hour Kit

For getting a wood fire started, these round, pine-scented sawdust-and-wax pellets can be used in place of tinder or kindling. They can be stored indefinitely in your 72-Hour Kit, and are a nonexplosive fuel source. They work with just about any type of fuel, including wood, coal, and charcoal. This economy-size box holds one hundred fire-starting nuggets.

WOOD PRODUCTS 9910 FATWOOD BOX

$$ | Amazon.com | 72-Hour Kit

Any dry wood will do to build a fire, but this ten-pound case of pine with a handle makes the wood easier to transport in your vehicle for an evacuation, and easier to stack with other stored supplies for your 72-Hour Kit at home. The wood is cut from pine stumps that have a high concentration of resin, which means it can be started with a match, even when it's wet—so it's also handy as a fire starter.

MAGNESIUM FIRE STARTER

$ | Amazon.com | 72-Hour Kit

Just in case all your matches get wet (or used), this can be a good backup. It works on the same principle as flint and steel: by creating a spark. The tool consists of a small block of magnesium that is waterproof and fireproof in its solid form. Using an attached serrated blade, you scrape some magnesium shavings into a little pile on something that catches fire easily, such as fibers, paper, moss, or kindling. Then you scrape the small flint rod to create a spark. The fire generated as the magnesium shavings burn is extremely hot and will ignite even damp kindling. All the parts fold together onto a package just 2.75 inches long.

CREOSOTE SWEEPING LOG FOR FIREPLACES (PACK OF 2)

$$ | Amazon.com | 72-Hour Kit

This is not a replacement for getting your chimney swept professionally, but in a pinch it can help reduce the dangerous build-up that causes chimney and attic fires. You place the log in your fireplace as you would any other log, but by itself. You light it, letting it burn by itself until the fuel has been exhausted. The chemicals embedded in the product work to eliminate some of the creosote that lines your chimney, although not all of it.

STOVE IN A CAN

$$ | Amazon.com | 72-Hour Kit

This little stove kit comes in a half-gallon can and contains four fuel cells, a fuel ring, a cooking ring, and waterproof matches. The fuel is nontoxic, water-resistant, and nonexplosive, and each cell burns for about one hour. Once the fuel is fully ignited, it boils water in about five minutes. It can be lit, used, extinguished (by placing the lid back on the can), and reused multiple times. The whole kit weighs 2.75 pounds and can be stored indefinitely. This is for outdoor use only.

FIRE SENSE 22-INCH FOLDING FIRE PIT

$$$ | Amazon.com | 72-Hour Kit

This little portable fireplace will accommodate any solid fuel, from wood to coal. It has a twenty-two-inch heat-resistant painted steel fire bowl and folding legs for portability. The heat-resistant mesh fire screen on top is a nice safety feature. It also includes a wood grate, a cooking grate, and a screen lift tool. The whole thing folds up into a mesh bag (which is included). The small size makes this great for throwing in the back of the car, and for those who have just a little bit of outdoor space—perhaps on the terrace of an apartment building.

COBRACO DIAMOND MESH FIRE PIT

$$$ | AvantGardenDecor.com | 72-Hour Kit

A pricier and bigger version of the portable fire pit is essentially a backyard fireplace on legs. This one is made of steel and measures 39 × 14 inches. The five-inch-wide table edge running around the fire pit could be useful for holding cooking items and might be used as an eating table if you are sheltering outdoors. It comes with a protective vinyl cover, and a wire mesh screen cover to protect you from flying embers. This is for outdoor use only.

On Alert

- Never use a camp stove or other flame-producing cooking device indoors.
- Pots and pans with nonstick surfaces are generally not safe to use when cooking over a grill or an open fire. They will burn.
- Store matches in a waterproof container along with a piece of fine sandpaper, so if the striking surface on the matchbox wears down, you can still light the matches.

CHAPTER 8

Communication and Navigation

CHAPTER 8

___

Communication
and Navigation

D isasters often separate family members. In the wake of the terrorist attacks on the World Trade Center on September 11, 2001, even while the ash was still graying the streets, heartrending notes and flyers were posted by people trying to find missing loved ones.

Disasters cripple all kinds of communication services. When Hurricane Katrina, one of the most destructive natural disasters in U.S. history, struck the Gulf Coast, it hit thirty-eight 911 call centers, disrupting local emergency services. More than three million phone lines in Louisiana, Mississippi, and Alabama were blown out, as were half the local radio stations and almost half the television stations.

In a crisis, we need to find our loved ones and to receive news and services. That's why it's critical when preparing for a disaster to have a family communication plan and an emergency radio—preferably the hand-crank type, so you are not dependent on batteries or electricity when neither may be available.

However you define family—whether it's just you or includes a spouse, children, pets, or a care attendant—you need to make a plan for communicating after a disaster. This plan will inform all members of your family what to do and how to do it. (Disabled people should have a medical alert button and register with the local police so they can get the help they may need after an emergency.)

If you have a smartphone, you should subscribe to e-mail and text alerts from your local Office of Emergency Management website (just type "Office of Emergency Management" and your town into any search engine). However, a hand-crank radio will work even when cell phone and other batteries die and the power is out all around you.

How to Create a Family Communication Plan

Discuss among family members the kinds of situations in which you might be separated during a disaster. Designate a relative or friend out of state who will agree to be the contact for family members, since it is likely to be easier to make a long-distance call than a local one during an emergency. Enter that person's name and number as your In Case of Emergency (ICE) contact in your cell phone. If necessary, get prepaid phone cards for public phones if any family member does not have a cell phone.

Make sure all family members who have cell phones know how to text, since texts are the recommended form of communication in a disaster (texting does not clog networks the way voice calls do).

Designate two meeting places where everyone can gather. The first one should be just outside your home, and the second should be outside your neighborhood. Make sure each child knows where the meeting place is and what to do while waiting for others to arrive. Go to the site as a family so children will know and remember.

Family members should always carry identification. Make sure to include maps of your local area in your Go-Bag, even if you know your way around. Roads may be closed or clogged and you might have to find alternate routes.

Carry Your Complete ID For each member of your family, including children, create an identification card to be kept in a wallet or school backpack. Templates are available for adults and children online at Ready.gov.

Supplies You'll Need: Communication and Navigation

AREA MAPS
$ | Amazon.com | Go-Bag | 72-Hour Kit

Collect maps of your town, region, and state—not the maps you use every day but maps you stash in your Go-Bag. MapQuest and Google Maps are good online resources for maps you can print in advance and place in a waterproof bag or pouch. For simplified maps of a downtown area that a child might easily follow, try your local chamber of commerce. Make sure the simplified map has the streets leading to your out-of-neighborhood meeting place, as designated in your family communication plan. You can even mark the meeting place on the map.

KWIK TEK DRY PAK MULTI-PURPOSE CASE
$ | Amazon.com | Go-Bag | 72-Hour Kit

You can buy a waterproof map case, but this product is more versatile for your Go-Bag. It's an inexpensive waterproof pouch made of clear vinyl that can be used to store maps and other items to keep them dry. The pouch measures 9 × 12 inches—big enough for maps, your ID, a cell phone, and a small flashlight. It has a yellow sealing clip and two D-rings at the bottom.

GARMIN ETREX 10 WORLDWIDE HANDHELD GPS NAVIGATOR
$$$ | Amazon.com | Go-Bag | 72-Hour Kit

If a technological device gives you more navigational comfort than maps, this handheld GPS is rainproof (not submersion proof). The 2.2-inch display is preloaded with a worldwide relief map and can accommodate topographic, marine, and road maps. It uses two AA batteries (not included) that will run it for twenty hours.

MILITARY PRISMATIC SIGHTING COMPASS

$ | Amazon.com | Go-Bag | 72-Hour Kit

A compass is an outdoor survival necessity. It shows the way to go. You can find a cheaper plastic model, or pay much more for style, better materials, and color, but in a disaster, this compass will serve your needs just fine. This basic military-style (made in China) compass comes with a water-resistant nylon pouch. You can get your bearings outdoors by seeing where you are heading—north, east, south, or west—and set out in the correct direction if you are using a map.

BUSHNELL GPS BACKTRACK PERSONAL LOCATOR

$$$ | Amazon.com | Go-Bag | 72-Hour Kit

This tracking device might be used during an emergency in which you have to hike or walk into areas unknown. If, for example, you had to park your car on the side of the road and walk some distance into a wooded area, you could program the device to lead you back to your car. It has a GPS receiver and a self-calibrating digital compass. It can store and locate up to three locations. It runs on two AAA batteries (not included) and comes in five colors.

MOTOROLA MH230R 23-MILE RANGE 22-CHANNEL FRS/GMRS TWO-WAY RADIOS

$$$ | Amazon.com | Go-Bag | 72-Hour Kit

Not essential to your Go-Bag but possibly useful for keeping in touch out-doors during a disaster, this pair of two-way radios, or walkie-talkies, has a range of twenty-three miles. We chose this model because it's waterproof. It runs ten hours on alkaline batteries or eight hours on rechargeable batteries (which are included). It also receives eleven weather channels. This set comes in a three-pack, too.

RED CROSS FRX1 ETON EMERGENCY RADIO
$$ | RedCrossStore.org | Go-Bag | 72-Hour Kit

There are three emergency radios the Red Cross not only approves but also sells, all manufactured by Eton. A basic radio is only twenty-five dollars, but as you get more features, the price goes up. On this most basic model, you crank the hand turbine for sixty seconds and get fifteen to twenty minutes of radio play. This is an AM, FM, and NOAA (National Oceanic and Atmospheric Administration) weather band analog radio. It also has a built-in hand-crank LED flashlight and it glows in the dark. There is an internal rechargeable nickel–metal hydride battery, and a headphone jack.

RED CROSS FRX2 ETON EMERGENCY RADIO
$$ | RedCrossStore.org | Go-Bag | 72-Hour Kit

Crank the hand turbine for sixty seconds and you'll get about fifteen minutes of radio play. This is an AM, FM, and NOAA weather band analog radio. It has an internal rechargeable nickel–metal hydride battery and a USB port, so you can charge it by plugging it into your computer. You can also use the USB port and the hand crank to recharge your cell phone. This radio can also be charged by plugging it into a wall outlet, or by using the built-in solar charger (if it's sunny). It includes an LED flashlight, a headphone jack, and glow-in-the-dark housing.

RED CROSS FRX3 ETON EMERGENCY RADIO

$$$ | RedCrossStore.org | Go-Bag | 72-Hour Kit

On this hand-crank radio you can get all seven NOAA weather band radio broadcasts, plus AM and FM radio with a digital display. This radio has an internal rechargeable nickel–metal hydride battery and an option to power it with AAA batteries. There's a DC power input with a mini-USB cable included, plus an AUX input to play music from an external MP3 player, along with a headphone jack. This model includes a three-LED flashlight and a red LED flashing beacon, plus an alarm clock.

ETON RED CROSS CLIPRAY FLASHLIGHT

$ | ShopEtonCorp.com | Go-Bag | 72-Hour Kit

While this is marketed as a flashlight, perhaps its best feature is that it doubles as a cell phone charger. In an emergency, you can't rely on electrical power, so having a hand-crank charger may be a lifesaver. Both the USB cell phone charger and the LED light run on a hand crank. This little device weighs just under five ounces and measures 2.25 × 6 × 1.25 inches. The Clipray flashlight is not an intrinsically safe flashlight, so don't use it if you suspect gas leaks.

SOLAR POWERED BACKUP BATTERY AND CHARGER FOR PORTABLE DEVICES

$$ | Amazon.com | Go-Bag | 72-Hour Kit

This solar charger uses the sun to charge up its internal lithium-ion battery. It then doles out the power to any USB-powered device. It can charge your phone battery only up to 50 percent, but in an extended power outage when you can't plug in your charger, this could be a real treasure in your 72-Hour Kit. It comes with a USB cable, suction cups so you can attach it to a home or car window to catch rays, and a little adjustable LED light.

How to Extend the Life of Your Cell Phone Battery

In a disaster, however much you might be tempted to follow the disaster on your smartphone, or to play diverting games or watch movies—don't. You need to leave communication lines unclogged by entertainment uses so that 911 emergency calls can get through. Your usual cell phone habits could also shorten the battery life of your phone.

Extend cell phone battery life by closing unused apps that draw power and by putting your phone in airplane mode. Because the phone won't search for a network signal in airplane mode, the battery lasts longer. If necessary, you can get news alerts on your car radio, but unless you are on the road, we recommend an emergency radio.

Keep extra batteries and phone chargers ready for a power outage. When your phone battery runs out, recharge your phone in the car, if possible—but beware of carbon monoxide fumes from the car exhaust if you have the car in an attached garage! Do not run the car unless the house is sealed from these fumes and the garage is well ventilated.

Subscribe to Alert Services Before a Disaster Strikes Most communities now have systems that will send an instant text message or e-mail to give you updates about bad weather, local emergencies, and other life-threatening warnings. Visit your local Office of Emergency Management website and sign up. Smartphone apps are another way technology can help keep you up to date, or save a life. There are numerous apps dedicated to helping people, and pets, during an emergency. Popular apps include everything from an Emergency Radio app that allows you to have the emergency frequencies of almost all the major police and emergency departments as well as air traffic control to Pocket First Aid, an app that informs you of procedures for saving a person's life.

On Alert

- During an emergency, so many people trying to use their wireless and land-based telephones at the same time creates network congestion. To free up network space for emergency communications (and preserve your phone battery), use texting to contact others, and limit your non-emergency calls.
- If you are redialing on a mobile phone, wait ten seconds after each call before dialing again, or you will contribute to clogging the network.
- If you must evacuate, forward your land line number to your wireless number to get incoming calls.

CHAPTER 9

Hygiene and Sanitation

CHAPTER 9

Hygiene and Sanitation

Disasters make hygiene complicated. Broken sewer lines can contaminate drinking water and expose you directly to untreated waste products. Flooding can pose one of the greatest dangers to hygiene because floodwaters, which can contain raw sewage, among other threats, disperse pathogens. The more than 150,000 people who remained in the New Orleans area when Hurricane Katrina made landfall were not only at risk from injury or drowning, but they were exposed during and after the hurricane to pathogens brought by floodwaters. Every surface of your home must be sanitized after a flood. And you should not drink the local water supply until it has been tested and declared safe.

Short of these serious health threats, disaster can mean no hot water for showers or baths, or no water, period. Those who have prepared will have some stored water and the right kind of bleach to purify water (see chapter 2 for more on safe water).

For first aid emergencies tended to at home or the workplace, your emergency kit will contain gloves, a mask, and plenty of hand sanitizer so that hygiene can be maintained while dressing wounds or tending to the sick (see chapter 5 for more on first aid). And, of course, use only purified water to clean a wound.

Finally, hygiene in a disaster might mean human waste must be collected and disposed of safely to prevent the spread of disease. Human waste must be buried in a secure location or stored in tightly sealed heavy plastic containers until proper disposal becomes available.

Bleach It All Away   The first recorded use of chlorine bleach as a medical disinfectant was in Austria in 1847 at the Vienna General Hospital. The right kind of chlorine bleach is a practical and effective disinfectant—the right kind means it must contain at least 5.25 percent hypochlorite as the main ingredient.

The food-processing industry uses chlorine bleach to kill hazardous bacteria such as listeria, salmonella, and E. coli on equipment. Sodium hypochlorite added to water kills the waterborne organisms that cause typhoid fever and cholera. Chlorine bleach can also kill dangerous bacteria and viruses on surfaces. It is especially valuable as a disinfectant because pathogens are not able to develop immunity against it, as they have done against certain drugs.

Supplies You'll Need: Hygiene and Sanitation

TOILET PAPER AND RELATED PRODUCTS
$$ | Amazon.com | 72-Hour Kit

The most difficult thing about storing toilet paper for your kit is finding room for it and making sure you don't use it up before an emergency! Buy the amount your family would use in three to seven days, then bag it (or even double bag it) in plastic for extra assurance that it stays dry. For infants and others with medical conditions who require commercially packaged cleansing wipes, make sure you stock up on those as well. And don't forget products for menstruating women and girls. Set aside a supply of plastic disposal bags with twist ties, for proper waste management during an emergency.

SCENSIBLES PERSONAL DISPOSAL BAGS FOR SANITARY PADS AND TAMPONS

$ | Amazon.com | 72-Hour Kit

These biodegradable bags for the sanitary disposal of tampons and sanitary pads come fifty to a box. They are lightly scented, measure 3.5 × 9.75 × 2 inches, and have a built-in tie handle closure. Antimicrobial agents are built right into the bag to inhibit the growth of bacteria.

GERBER 22-41578 GORGE FOLDING SHOVEL

$$ | Amazon.com | 72-Hour Kit

For those fortunate enough to have a yard in which to dig a latrine, this folding shovel will work fine. But if your yard has been flooded, use the home-made emergency latrine described at the end of this chapter, or purchase one. This folding shovel can be used for mud and snow removal, as well as latrine digging and other emergency tasks. It has a padded nylon handle and comes in a handy bag.

COMFORT BATH PERSONAL CLEANSING, ULTRA-THICK DISPOSABLE WASHCLOTHS

$ | Amazon.com | 72-Hour Kit

There won't always be water for bathing. When you're conserving all water for drinking, use these no-rinse washcloths that come in a box of four packages, each containing eight disposable washcloths. They contain a rinse-free cleanser, plus aloe vera and vitamin E.

PURELL SANITIZING HAND WIPES

$ | Amazon.com | 72-Hour Kit

Must-haves, these economically priced wipes earned the highest ratings from *Consumer Reports* for effectively sanitizing hands without water. Each wipe is 5 × 7 inches. This is the large box of one hundred individually wrapped wipes. Wipes that are not individually wrapped and come in large pop-up containers can dry out after the container is opened. Hand sanitizer also comes in pump bottles, and those work fine, but a wipe is easier if you want to clean your face or your children.

RELIANCE LUGGABLE LOO TOILET SEAT AND BUCKET

$$ | AustinKayak.com | 72-Hour Kit

This bucket provides a place to eliminate and contain the waste when using an outdoor latrine is not feasible. More than just a five-gallon bucket, it is topped with a round seat that users report is very sturdy. The bucket must still be lined with a plastic bag (not included) for proper waste disposal.

CLEANWASTE GO ANYWHERE PORTABLE TOILET

$$$ | Amazon.com | 72-Hour Kit

This toilet folds up into a 5 × 19 × 14–inch package, and unfolds to roughly the same height and seat dimensions as a standard toilet. A removable mesh holder supports a waste-collection bag, so you can use plastic bags as waste receptacles. It comes with three biodegradable waste-collection bags that use eco-friendly powder to turn liquid waste into an odorless solid. A lid protects the seat when not in use and removes for stability under the legs when you set up on soft ground. The toilet weighs seven pounds and will support up to 250 pounds.

TOILET SEAT

$ | Amazon.com | 72-Hour Kit

More economical than buying a whole portable toilet kit, you can just buy this toilet lid and use it with the homemade emergency toilet described at the end of this chapter. The molded seat and lid snaps onto a five-gallon bucket container.

CLEANWASTE BULK POO POWDER

$$$ | Amazon.com | 72-Hour Kit

Although it is kind of pricey, this nontoxic powder makes it possible to extend the life of your waste collection bags used for portable latrines. It works a bit like clumping kitty litter, turning liquid waste into a solid. One scoop of powder treats up to twenty-one ounces of liquid or solid waste, preventing backsplash and making the waste easier to transport cleanly. This canister contains 120 scoops of waste treatment and a measuring scoop.

WORLD'S BEST CAT LITTER MULTIPLE CAT CLUMPING FORMULA

$$ | PetFoodDirect.com | 72-Hour Kit

Clumping kitty litter will do much the same job as Poo Powder—capturing liquid waste and clumping it into a ball of litter, which can help control the waste in your emergency latrine. This litter gets high marks for clumping, is made from corn, and is biodegradable. It also comes in a lavender-scented variety. Get the twenty-eight-pound bag multiple-cat formula—a single cat's liquid output is far less than a human's.

STANSPORT CABANA PRIVACY SHELTER TENT

$$$ | Amazon.com | 72-Hour Kit

While a tarp and some nylon cord can create privacy, this free-standing privy tent is sturdier and offers more complete privacy. It is supported by fiberglass poles with sleeved guides and stake-out rings for windy areas. The tent itself is nylon with screened windows and vents designed to block the view but allow ventilation, and it has a zippered door. A mesh storage pocket inside gives you a place to put the toilet paper. It measures 7 × 4 × 4 feet when set up.

How to Make an Emergency Toilet

SUPPLIES

- 5-gallon bucket
- Heavy-duty garbage bags with twist ties
- Liquid antibacterial soap and/or towelettes
- Plain household bleach (at least 5.25 percent hypochlorite)
- Substrate: sand, sawdust, clumping kitty litter, or shredded newspapers
- Toilet paper

ASSEMBLING THE TOILET AND HYGIENIC "FLUSH" LIQUID

- Line the bucket with a garbage bag.
- Put substrate in the bottom to prevent splashes and absorb liquids.
- Make a solution of bleach—1 part bleach to 10 parts water (1:10).

USING THE BUCKET

- After using the bucket, pour some bleach solution into the bucket (to help prevent air-born contaminants from spreading).
- Use a twist tie to close the toilet bag between uses.
- Wash your hands with liquid soap or use a sanitizing wipe to cleanse.

MANAGING THE WASTE

- Change the toilet bags daily.
- Do not put the waste-filled bags in the garbage.
- Collect the bags until there is a means of properly disposing of them after the disaster.

Before Everything Hits the Fan Septic systems and community wastewater systems play vital roles in sanitation and disease prevention by removing harmful viruses, bacteria, and parasites. If damaged during an emergency such as a flood, hurricane, or earthquake, these systems can become very dangerous. This damage can lead to contamination of the environment and drinking water supply and result in an increased risk for disease. Be sure to prepare your septic system before an emergency by sealing the utility hole and/or inspection ports to keep excess water out of the septic tank. To prevent your septic tank from collapsing or floating, make sure it is at least half full. If your septic system requires electricity, turn off the pump at the circuit box before the area floods and waterproof all electrical connections to avoid electrical shock or damage to wiring, pumps, and the electrical system.

During a state of emergency, reduce the amount of water used by limiting toilet flushing, dishwashing, washing clothes, and showering and be sure to avoid contact with any standing water that may contain sewage.

On Alert

- If the sewer lines are not working, line your toilet bowl with a plastic bag and use the toilet as instructed for an emergency toilet. Your own toilet will still be more comfortable because it is private and has a seat.

- Keep your latrine and bags of stored latrine products far away from any room or surface where food is being prepared. Never prepare or store food where human waste or blood is or has been present. Do not consume foods that have been exposed to either substance. *When in doubt, throw it out.*

- Do not place human waste in the garbage. After the disaster there will be a way to properly dispose of all accumulated waste bags. Ask for this information if it is not provided.

- Washing your hands with soap and water is the best way to reduce the number of germs on them, and prevent germs from spreading. If your tap water is not safe to use, wash your hands with soap and water that has been boiled or disinfected. If soap and water are not available, use an alcohol-based hand sanitizer that contains at least 60 percent alcohol. Alcohol-based hand sanitizers can quickly reduce the number of germs on hands in some situations, but sanitizers do not eliminate all types of germs.

CHAPTER 10

Heating and Cooling

CHAPTER 10

▬

Heating and Cooling

D isasters strike in any season, so you must prepare for extremes of both heat and cold. This means keeping your body and your sheltering places cool or warm, even if there is no electricity, or the gas or oil heater is not working.

Remember that the very young and the elderly are less able to regulate their body temperature. Some pets overheat more quickly than we do (dogs, for example), and some chill at higher temperatures than we do (including reptiles, tropical birds, and small rodents such as gerbils and hamsters). All will need special consideration.

Cooling Down

The Centers for Disease Control and Prevention (CDC) takes hot weather very seriously and urges us to prepare for extreme heat, which can be life-threatening. For both comfort and protection, wear loose clothing that is lightweight and light colored to reflect heat rather than absorb it.

It's important to keep hydrated, and the best beverage for that is water. Alcohol and sugary drinks rob the body of fluids at a time when water in the body is crucial. Drink plain water, and drink it continuously during hot weather events, even before you are thirsty. If you're doing anything physical, drink two to four glasses of water an hour. Keep in mind that highly iced drinks can cramp your stomach.

If you're sweating a lot, you'll also need to replace electrolytes. Electrolytes are essential salts, including potassium and sodium, that are necessary for the metabolism of all the cells in your body. You can replace them by drinking a prepared electrolyte replacement solution such as Pedialyte (sports drinks tend to have a lot of sugar) or by taking an electrolyte supplement. They are available as gel, tablets, powder packs, and strips; popular brands include Gatorade. Potassium-rich foods include bananas, raisins, and dried apricots. Half a teaspoon of iodized salt in a glass of water will replace lost sodium.

In a hot-weather disaster, if your air conditioner doesn't work or you don't have air-conditioning, visit a public library or mall that does have air-conditioning. Some areas have cooling shelters for extreme heat events. Call your local health department to see if there are any heat-relief shelters in your area.

If you cannot get to a place with air-conditioning but you do have electricity, an electric fan will not provide enough cooling to prevent heat-related illness. Add some ice water to a tray set in front of your fan to step up the cooling factor. If you have a basement, stay there because it tends to be cooler below ground level. Keep your body damp with a handheld spray bottle to facilitate cooling evaporation.

If you have in your care—or in your neighborhood—those who are at greater risk for heat illness, check up on them and make sure they are being watched carefully for symptoms of heat stroke or heat exhaustion. The most vulnerable individuals are babies and children, adults over sixty-five, those with mental illness, and the physically ill, particularly those with high blood pressure or heart disease.

If you can, stay indoors. It you must go outside, stay in the shade or set up a temporary lean-to (see chapter 4) to create shade. Wear a wide-brimmed hat and a broad-spectrum sunscreen. Rest in the shade as often as you can, and drink plenty of water.

Summer in the City Cities and suburbs are between one and ten degrees hotter than rural areas during a heat wave, because of the preponderance of heat-absorbing surfaces in urban landscapes, such as asphalt and dark-colored roofing materials. Cities also tend to have a higher concentration of cars and other heat-producing machines. The countryside can get pretty hot, but it's a cooler place when it's sweltering.

Supplies You'll Need: Cooling

CDN DTQ450X PROACCURATE QUICK-READ THERMOMETER
$ | Amazon.com | 72-Hour Kit

Keeping your food safe during power outages requires a good thermometer, and the digital type produces a reading in seconds rather than minutes. Bacteria multiply best in a warm, moist environment. Perishable food that is between the temperatures of 40 and 140 degrees Fahrenheit provides an ideal environment, and bacteria will multiply at an exponential rate. After two hours in this temperature zone, there will be too much bacteria, and the food needs to be thrown out—even if it looks and smells okay. That is why it is so important to keep foods cold and to follow the adage: when in doubt, throw it out!

KAFKA'S KOOL TIE
$ | REI.com | 72-Hour Kit

This lightweight scarf is filled with polymer crystals. You soak the tie in water (it need not be cold water) and the crystals hydrate. Then wrap the tie around your neck and the crystals slowly release their water, facilitating evaporation on your skin. Evaporation is the body's natural cooling mechanism. Kool Tie is soft and comes in eleven colorful prints.

INSTANT COLD COMPRESSES
$$ | First-Aid-Product.com | 72-Hour Kit

Cold chemical compresses are typically sold for use on sports injuries, such as sprains and strains. But a quick shot of cold on the back of the neck, armpit, groin, or core area can also help revive a person or animal who is overheated. This case of twenty 6 × 9–inch instant cold compresses requires no pre-chilling—just twist to mix the chemicals sealed inside, and you get instant cold. They are disposable and can't be reused. They stay cold about thirty minutes.

COOL ON THE GO
$$ | Amazon.com | 72-Hour Kit

While there are more economical ways to cool yourself, including self-misting with a spray bottle of ice water, or hanging a wet sheet in the window, these thing don't travel well. This little air conditioner clips to your shirt or belt or the baby's stroller, and provides a continuous stream of cool air. It uses four AA batteries (not included) that will run it for about five hours, but it also has a USB port (cable included) so you can run it off another power source. It weighs a little over one pound.

 Preparing for Extreme Heat

- Apply light-colored paint to asphalt; use light paving materials for new projects.

- Get a food thermometer for power outages.

- Install air-conditioning and locate public cooling shelters in your area.

- Insulate your home.

- Plant trees to create cooling shade near the house.

- Use light or white roofing materials.

On Alert

- Don't top off gas tanks when fueling in hot weather. Vapors from spilled gas can ignite.
- Never leave any person or animal unattended in a car or a pool.
- Reduce your electricity consumption in extreme heat to help prevent brownouts or blackouts. This is not the time to run the dishwasher or do laundry.

Warming Up

Prepare for cold weather disasters with warm clothing and bedding. Dressing in loose, insulating layers will trap your body heat and keep you comfortable in cold weather. If the heat is out in your home, dry sleeping bags or dry blankets create extra insulating layers for warmth. Eating healthful meals will also help your body stay warmer.

If you live in an apartment or a rental where you can't add insulation, placing blankets over windows and at the cracks under doors helps keep heat inside.

If the heat is out, gather everyone together in a single room of the home, preferably a small one, and do everything you can to warm up just that room. The collective body heat will also help. Use the fireplace to stay warm, but never use a charcoal or gas grill indoors and never run a generator indoors because you will be exposed to deadly fumes and carbon monoxide. And never leave a candle unattended.

For injured persons and others at risk from cold out of doors, a space blanket will help prevent hypothermia. The space blanket is wrapped like a tube around the victim and duct-taped closed with enough room for a layer of insulating air to remain between the injured person and the blanket. This creates a cocoon in which the person's body warmth is trapped, and the resulting humidity in the tube slows the perspiration process, so there is less body heat lost. In emergency situations, a space blanket can be used to reflect light and show your position to rescuers. It can also be used to improvise an emergency shelter (see chapter 4).

Preparing for Extreme Cold

- Add snow shovels and sand to your 72-Hour Kit.

- Clean chimneys and fireplaces at least annually.

- Insulate water pipes and the water heater.

- Insulate your home.

- Put warm clothing and bedding in your 72-Hour Kit.

- Store extra fuel for your fireplace or wood stove.

Supplies You'll Need: Heating

SUNCAST SF1850 22-INCH BIG SCOOP SNOW SHOVEL WITH WEAR STRIP
$$$ | Amazon.com | 72-Hour Kit

The scoop design of this shovel spreads the weight of the shovel's contents evenly across both your arms, making shoveling easier on your back. It is five feet long overall, and has a 22 × 28 × 5–inch graphite blade. The steel handle is collapsible for easier storage.

QUANTUM HEAT PACK
$–$$$ | QuantumHeatPacks.com | 72-Hour Kit

If a Mylar blanket just will not warm a person up, the situation can get dangerous. A heat pack can provide quick warming. These packs come in a variety of shapes and sizes, and need no power. When activated by pressing a button, the chemicals inside begin to crystallize, producing heat up to 130 degrees within a few seconds. Depending on the size of the heat pack, they can stay hot from one to three hours. These heat packs are reusable—you simmer them in hot water to reverse the chemical process, and they're ready to go.

THERMWELL SP57/11C WATER HEATER BLANKET
$$ | Amazon.com

Keeping your water heater insulated is a good idea in all weather because it lowers your energy bill. During disasters, a thermal covering over the tank can prevent freezing, which will ensure that you have an emergency water supply in your house if you are trapped there in freezing weather and water is not coming out of the tap. You can pay a lot more or a bit less for a water heater blanket. Or you can make one yourself by cutting insulation pieces to fit the tank and pipes, according to *Popular Mechanics* (PopularMechanics.com)— although this is not recommend unless you know how to work with insulation and custom fitting it.

SPACE ALL-WEATHER BLANKET
$ | REI.com | Go-Bag | 72-Hour Kit

More expensive yet more substantial than the plain Mylar blankets that typically stock first aid kits, this space blanket has grommets and a lining so it is less likely to tear and more likely to withstand heavier use in rough conditions. Having grommets allows you to feed a nylon cord through the blanket without tearing it. You can then tie the cord to a tree, if you plan to use the blanket for shelter or for protection from the weather (see chapter 4).

EMERGENCY ZONE BRAND EMERGENCY SLEEPING BAG, SURVIVAL BAG, REFLECTIVE BLANKET

$ | Amazon.com | Go-Bag | 72-Hour Kit | Car Kit

Good for your Go-Bag in cold weather or in your auto kit, this bag-shaped space blanket can help retain body heat and prevent hypothermia in an emergency. You can make the same thing with space blankets and duct tape, but this saves time. All you have to do to use it is open it up and slip yourself (or someone else) inside. There are little vents so you don't get too hot, but for the most part, the bag traps your body heat and keeps you warm.

Snuggle Up Adults over sixty-five are less efficient at creating body heat, and infants lose body heat very quickly and can't create it by shivering, as adults can. You can keep babies—and even adults—warm in cold temperatures by sharing your body heat.

On Alert

- If you have electricity, small space heaters can be a fire hazard, even though it is hard to imagine a fire danger when the world appears frozen. Never leave the space heater on when no one is in the room, don't sleep near it (blankets can catch fire), and don't run heater cords under carpets.
- Set up a secure outdoor box so if there's a power outage, you can store food in the outdoor cold without danger of animals getting it.

CHAPTER 11

Protection

CHAPTER 11

Protection

The first step and the most important preparation for any disaster is mental: imagine what your world would be like if nothing was operating as usual, then gather as much information as you can *before* running out to buy five years' worth of emergency rations. Sometimes your imagination can run wild, and gathering solid information will ground your preparations in reality.

Knowing what to expect and how to be ready for it will help you fight the worst effect of any disaster: fear. Fear is toxic to you and to others. It can slow livesaving tasks and infect others around you. Panicked people go from being part of the solution to part of the problem. The better prepared you are, the less fear will weaken your reasoning, reliability, and performance.

Unfortunately, terrified animals are incapable of the mental discipline of remaining calm in a disaster. Fearful animals will act in unpredictable ways, and sometimes endanger those around them. To protect yourself and your family from animals—including family pets—who "freak out" in a disaster, we recommend fully preparing your own animals for a coming disaster (see chapter 12). Beyond that, we recommend a few supplies, listed in this chapter.

Some Americans feel owning a firearm is their best protection. If you own or plan to purchase a firearm, be sure to adhere to your state's laws. But if you don't feel you need a gun in normal times, a disaster is no time to get one. Water and food will do far more good. People do not typically attack when possessed by excessive fear, nor do they want to harm you or take your things. Law enforcement officials do not recommend guns for protection. Further, the most commonly reported aberration in normal human behavior during disasters is a marked increase in kindness, generosity, and heroism.

How to Protect Your Home

Law enforcement officials say that in a disaster people may loot stores and other commercial businesses, but usually do not loot homes. Of course, there are always exceptions—typically empty homes, not ones where the occupants are sheltering in place. If you must evacuate, there are ways to protect your property before you go—both from the effects of whatever disaster is coming and from looters.

How to best protect your property from disaster depends on what type of disaster is on its way. For FEMA's advice on the steps to take to protect your home before a natural disaster, go to Ready.gov, select the specific disaster you are concerned about, then click on the "before" tab.

Home protection experts also offer some simple tips on how to best protect your property from people with bad intentions. To start, move your good stuff—the flat-screen TV, computers, and other temptations—away from the windows where people can see them from the outside.

Secure your doors by having the kind that are made of solid steel or sturdy hardwood, with no glass panels. Invest in high-end dead bolt locks for all your doors. Make sure the lock has a bolt that extends at least an inch into the door frame and that the strike plate (the metal piece that mounts on the door frame and holds the bolt) is attached with screws that are at least three inches long. Installing a new dead bolt is a skilled task, so it's best to leave it to a locksmith.

Securing your windows starts by making them hard to get to. Plant rose bushes or other thorny plants in front of all your ground-level windows. Would-be looters are less likely to want to prick and scratch their way through your bushes and more likely to move on to some other home that's easier to approach. You can also cover your windows from the inside with a big protective sheet of plastic, held up with duct tape. It's not a iron-tight security measure, but it does make the windows harder to break—making your home a less attractive target. Also consider security window film (see the following protection supplies).

Secure sliding glass doors and sliding windows by cutting dowels and setting them in the track. There are many small, inexpensive devices, from locking pins to wedges, that make it very difficult for an intruder to pry open double-hung windows. Home improvement stores typically carry a range of these products. Consider security bars on basement-level windows as well. You can even get bars that release quickly (from the inside) in case you need an emergency exit.

If you live in an area that is prone to hurricanes or high winds, storm shutters can keep your home safe from both human and weather threats. There are many styles of storm shutters to choose from: accordion-style, the kind that swing open and closed, and the kind that roll into place. They're not cheap, though (the cost depends on the style you choose and the manufacturer), and you will likely have to also pay for professional installation.

Mental Preparation "Preparing the Mind for Disaster: How to Replace Negative Thinking with Positive Thoughts" is one example of the excellent resources available on the Internet about preparing yourself mentally for a disaster. This article, by psychology writer Joanna Fishman, gives you a list of normal human responses to trauma as well as some helpful mental exercises to do before and during a disaster: http://www.howtosurvivestuff.com/survival-plan/preparing-the-mind-for-disaster-how-to-replace-negative-thinking-with-positive-thoughts. Since it is unlikely you will have Internet access during a disaster, print out a copy and store it with your other emergency documents.

Supplies You'll Need: Protection

DOG DAZER II ULTRASONIC DOG DETERRENT
$$ | Amazon.com | 72-Hour Kit

As discussed in this chapter, animals are the ones who can lose their manners and self-control during disasters because they don't know what is going on. This does not mean they will attack or bite you, but the possibility does exist. If there is a possibility you might encounter confused and excited dogs who seem ready to bite during or after an emergency, this device emits a high-pitched sound that will deter a dog that is within fifteen to twenty feet of you.

SABRE RED POLICE-STRENGTH PEPPER SPRAY
$ | Amazon.com | 72-Hour Kit

Pepper spray is a very serious weapon for self-defense, and if used improperly can do lasting damage to the eyes, nose, and throat of animals—including humans. If you feel threatened during a disaster scenario and the threat is very close, the briefest shot of this liquid hot pepper spray will keep the animal from coming near you. However, make sure you are not in close range of any other humans, because even at a distance this can cause severe burning in the lungs and eyes.

MILK BONE BISCUITS FOR SMALL AND MEDIUM DOGS
$ | Amazon.com | 72-Hour Kit

Experts who advise postal workers have suggested that keeping a few dog treats handy can sometimes be the best solution to a barking dog who seems unfriendly. They might also be handy to treat your own pet during an emergency, especially if the animal needs bribing to enter a pet carrier. If the dog seems dangerous, it is advisable to attempt bribing with food before resorting to an ear-splitting whistle or a shot of pepper spray.

SMALL ANIMAL TRAPS BY TRU-CATCH TRAPS
$$–$$$ | TruCatchTraps.com | 72-Hour Kit

A metal trap might be useful if you need to humanely trap your own or others' frantic animals after a disaster. By using food to lure the animal into the cage-like trap, you can secure the beast easily with the drop-down locking doors, then carry the animal to a shelter or perhaps just into the garage where it can remain restrained from hurting anyone. The trap will also provide a sense of protection for the animal itself. These traps come in a variety of sizes and styles.

BUDDYBAR DOOR JAMMER
$$$ | Amazon.com | 72-Hour Kit

This works just like those scenes in the movies where someone wedges a chair under the doorknob to prevent the door from being kicked in. This more secure bar also wedges between the doorknob and the door. It's recommended by security experts because it has no plastic parts and is made of powder-coated heavy-duty steel. It extends from thirty-six to fifty-one inches, and works on carpet, tile, concrete, and wood floors. Of course, you must set a door jammer from the inside, so this is for use when you are at home, and for your least secure doors when you are not at home.

3M SAFETY AND SECURITY WINDOW FILM
$$$$–$$$$$ | Solutions.3M.com

Security film is applied to the insides of windows. If something is thrown or blown into a window and it breaks, the film is designed to help hold the glass fragments together, reducing potential injury from flying glass and creating a stronger barrier that may slow down looters to the point where they may decide to pick an easier target. 3M makes window films in different thicknesses, and with or without sun control. Don't confuse this with tinted window films sold at home improvement stores that are designed purely for decoration or sun control and offer no security features. Security film should be professionally installed, and can be five to ten dollars per square foot, or more.

A PARADISE BUILT IN HELL: THE EXTRAORDINARY COMMUNITIES THAT ARISE IN DISASTER

$ | Amazon.com | 72-Hour Kit

To educate yourself about how little there is to fear from other humans during a disaster, this book by award-winning writer Rebecca Solnit (published by Viking Adult) chronicles the global patterns of human behavior during and after disasters. Solnit researched major disasters around the world, starting with the 1906 San Francisco earthquake and advancing to present times, uncovering evidence of communities that arose during emergency situations. She found that cooperation and kindness in times of disaster crosses all cultural and ethnic boundaries.

RESILIENT BY DESIGN: TRAIN YOUR MIND TO NAVIGATE LIFE'S OPPORTUNITIES

$ | Amazon.com | 72-Hour Kit

Part of what keeps people from establishing a healthful preparedness ethic is found in the unconscious perception of what psychologists call a victim role. To survive a disaster, people must change their thinking from weakness to strength, and adapt their self-perception to include a lack of fear. Written by disaster preparedness expert Ed Copp, this self-help book guides readers away from a victim role through perceptual shifts, moving them toward the ability to embrace change.

Animal Encounters

- **Cats** will most likely flee, so you typically don't need to worry about backing off. However, never corner a cat, because if the cat has no way to flee, it will fight.

- **Dogs** will not likely bother you unless you behave in a way they perceive as threatening. If a dog gets very stiff, holds up its tail, and begins to snarl at you while staring directly at you, there is danger. Showing teeth is a danger sign as well. Do not look the dog in the eyes, or the dog will think you are issuing a challenge. Turn so your body is sideways to the dog and back away. But don't run and don't turn away completely so your back is to the dog. Wherever you may be during or after a disaster (or today, even), don't run or walk fast past a dog you don't know. Act cool and calm, avoid turning your back on it, and you will avoid getting bitten.

- **Wild animals** are unlikely to approach you unless they are very hungry or have become trapped in your home. If you are sheltering out of doors, keep your food sealed in plastic bags and stored at least one hundred yards from where you sleep. Keep a very clean campsite, and dig trenches far away from camp (one hundred yards or more) for burying garbage, especially if you're camped in one area for an extended length of time.

- Do not corner wild animals or try to rescue them. Wild animals, first and foremost, want to get away from you, and may endanger themselves by dashing off into floodwaters or fire.

- Do not approach wild animals that have become trapped in your home. Snakes, opossums, raccoons, and other animals may seek refuge from floodwaters on upper levels of buildings, and have been known to remain after water recedes. If you come home after a flood and find animals in your home, open a window or provide another escape route and the animal will likely leave on its own. Do not attempt to capture or handle the animal. If the animal will not leave, close the door to the room it's in, open a window, and call your local animal control office or wildlife resource office.

- Do not attempt to move a dead animal. Animal carcasses can present serious health risks. Contact your local emergency management office or health department for help and instructions.

On Alert

- Knowing you are prepared is the best preparation you can have against being overwhelmed by fear.
- Check in on neighbors to make sure they are not scared. Companionship helps chase fears away and lowers stress levels during and after a disaster.
- Do not approach unfamiliar dogs or any other animals during or after a disaster. Keep children away from all animals during and after an emergency, unless you are certain the animal is both yours and is exhibiting normal behaviors. If you are bitten by any animal, seek immediate medical attention.

CHAPTER 12

Pet Safety

CHAPTER 12

Pet Safety

A nimals need and deserve as much care and comfort during an emergency as their owners—and this includes livestock as well as pets. We have to plan for the safety, food, and drinking water of our animals more even than for ourselves in an emergency, because they are incapable of anticipating or planning for a disaster. Animals, too, need a Go-Bag for evacuation, because if you evacuate, your animals must evacuate. As the Red Cross puts it, "If it is not safe for you to stay, it is not safe for them, either."

Part of your emergency preparedness is sympathy and an understanding of animal psychology. Animals can respond with panic to a disaster and exhibit behaviors that are out of character. Animals can sense when a severe weather event is about to occur, so be aware that your pets or livestock might try to isolate themselves in advance of a dangerous storm. Keep animals near you, and keep newspapers set aside in case a terrified pet becomes unable to control bodily eliminations. Other behaviors can occur as well. A laid-back dog can suddenly act aggressively or defensively. Not so unlike people, animals can respond to a traumatic event in surprising and even dangerous ways, and must be protected and soothed until they can behave calmly again.

Keep in mind that not all Red Cross and other evacuation shelters accept pets, so you must plan in advance for their safe boarding after a disaster. Ask your veterinarian for a list of boarding kennels and facilities. Ask your local animal shelter if they provide emergency shelter for pets. Identify hotels or motels outside of your immediate area that accept pets. Ask friends and relatives outside your immediate area if they would be willing to take your pets.

In case of a fire, all pet owners should have decals on their home windows—known as pet fire safety decals—that alert firefighters that animals are inside and at risk.

If you own livestock, after a severe weather event or earthquake, make sure that fencing—which may have been damaged by the disaster—is intact so fearful animals don't harm themselves or others.

Never Leave a Pet in the Car Never, ever leave your pet alone in a parked car. Even with the car windows partially open and even if you have parked in the shade, on a day when it's just in the 70s, the temperature inside a parked car can reach 120 degrees Fahrenheit. Your pet can quickly suffer brain damage or die from heatstroke or suffocation.

Pet Safety Before, During, and After a Disaster

BEFORE

- Observe now and become familiar with your pet's normal behavior.
- Keep your pet up-to-date with vaccinations and parasite control.
- Attach a secure tag with up-to-date contact information to your pet's collar.
- Consider microchipping all your animals (learn more at Pets.WebMD.com).
- Have an appropriate, secure crate, carrier, or trailer for every animal that can't be walked or led, and train the animal to use the carrier calmly.
- Store all carriers where you can easily and quickly get to them.
- If possible, train all your pets to respond to a few simple cues, including "come" and "stay."
- Include pets in family evacuation drills, so the ritual becomes familiar to the animal.
- Make a list of emergency animal shelters and homes of friends where your animal might stay.

DURING

- If you are aware in advance of danger, bring your animals indoors with you (or into their barn or shelter).
- Observe behavior changes, including disorientation, and keep animals under close and direct control.
- Keep animals of different species apart during a disaster, even if they are best of friends under normal circumstances.
- Keep smaller pets away from dogs and cats.

AFTER

- Immediately address behavior changes to keep animals and people safe.
- Allow animals outside only on a leash or lead.
- The danger of snakes is higher after a wet-weather or fire disaster, as is the danger of downed power lines.
- Be alert to the danger of spilled chemical products affecting the feet or noses of animals.
- Keep animals away from access to areas where there might be broken glass, twisted metal, or any other danger.
- Do not let animals drink floodwater or any other water sources that may be contaminated as a result of a disaster.
- Reintroduce food in small servings, gradually working up to full portions, if animals have been without food for a prolonged period of time.

Don't Leave Spot or Boo Behind When emergencies strike, some people may be tempted to leave their animals behind or let them go out on their own. During Hurricane Katrina, some 50,000 pets were left behind, by conservative estimates. Watch the volunteer rescue effort chronicled in the eye-opening film *Dark Water Rising: The Truth About Hurricane Katrina Animal Rescues*. Apart from guardianship being your responsibility to your pet, in a state of emergency pets can be as much of a necessary tool as power or food supplies. They can help you defend your home against intruders and take care of any rodent or insect problem that may arise.

Be Patient When disaster strikes familiar landmarks and smells might be gone, and your pet will probably be disoriented. It is not uncommon for pets to get lost, run away, or act out during a disaster. Make an effort to get your pets back into their normal routines as soon as possible. The stress of the situation will very likely cause behavioral problems. If these problems persist, or if your pet seems to be having any health problems, talk to your veterinarian.

Supplies You'll Need: Pet

PET SAFETY ALERT DECAL
Free | ASPCA.org

Place a pet alert sticker on the window near your front door so rescue personnel or firefighters will know that animals are inside. Indicate on this label what kinds of pets you have and how many, plus an emergency phone number. The ASPCA offers these decals for free.

TWO-DOOR TOP-LOAD KENNEL
$$ | DrsFosterSmith.com | Go-Bag | 72-Hour Kit

This carrier comes in small and medium, for both cats and small to midsized dogs. It's molded from rigid plastic and includes water and food bowls. It has both a steel wire front door and a top-loading door to easily load and unload your pet. Secure, spring-loaded latches enable you to open the kennel one-handed, minimizing the risk of escape. The top and bottom halves are secured with bolts and wing nuts, so the top can easily be removed. The carrier meets International Air Transport Association (IATA) and United States Department of Agriculture (USDA) air travel requirements.

POQUITO AVIAN HOTEL

$$$ | DrsFosterSmith.com | Go-Bag | 72-Hour Kit

This is a folding travel cage that knocks down and sets up with no nuts or bolts. The natural wood perch and handle is easy on your hand to carry and provides an out-of-the-cage play area. The cage has half-inch bar spacing and is made from sturdy iron covered in nontoxic lead-free and zinc-free powder-coated paint. It comes with two stainless steel cups with outside access and one cotton-rope perch. The cage is 18 × 14.25 × 15.25 inches inside.

NATURE'S MIRACLE DISPOSABLE LITTER BOX

$ | DrsFosterSmith.com | Go-Bag | 72-Hour Kit

This disposable cat litter box is made from 100 percent recycled paper. Its biodegradable design promotes air flow to help keep litter drier. It's sturdy enough that it will not leak, tear, or shred under normal use. Each box lasts up to four weeks. The front edge is slightly lower, making it easier for your cat to get in and out. The box comes in three-packs for the regular size (13 × 16 × 4.5 inches) or two-packs for the jumbo size (15 × 20.6 × 6 inches).

SOLVIT CAR SAFETY HARNESSES FOR DOGS

$$ | Amazon.com | Go-Bag | 72-Hour Kit

If you are going to evacuate by car, your dog might need a little help being well behaved during the car ride. This harness attaches to your car's rear seat belt and will keep your dog in the back seat while you drive. This is safer for the dog and safer for everyone in the car. The harness does not present any danger of strangulation because the restraint does not go around the dog's throat. The device comes in a range of sizes to accommodate all sizes of dogs.

SAFETY KATZ WALKING JACKET FOR CATS
$$ | JoyKatz.net | Go-Bag | 72-Hour Kit

Your cat may not appreciate being confined in a carrier for any length of time, but it may not be safe to let your kitty wander around in a shelter, outdoors, or sometimes even in the house, depending on your situation. Cats will slip out of a collar and leash, but a walking jacket is a safe and comfortable alternative. These walking jackets come in a very wide array of sizes and styles. All have a ring at the back to attach a leash, which is not included. (Be aware that you must train your cat to accept a walking jacket and leash before disaster strikes. For information on how to do that, visit ASPCA.com.)

EMERGENCY READY DELUXE PET FIRST AID KIT
$$ | ASPCA.com | Go-Bag | 72-Hour Kit

This animal first aid kit is sold by the ASPCA and was assembled with special attention to the injury and sickness needs of pets. Having an animal first aid kit is essential in a disaster because you don't want to draw on the supplies you have set aside for human use, and you want to have medications and topical products developed for animal use, along with dressings made to fit the bodies of furred creatures.

PET ID BARREL TAG ADDRESS LABEL TUBE
$ | Amazon.com | Go-Bag | 72-Hour Kit

Rather than an ID tag, this is a small, watertight, aluminum alloy tube that attaches to any ring on a pet's collar. The tube unscrews, and you put a piece of paper inside containing all your emergency contact information. The advantage is that if you are on the move, or simply not at home, you can keep changing the information inside the tube as needed.

VETRI-SCIENCE COMPOSURE CANINE BITE-SIZED CHEWS
$ | Amazon.com | Go-Bag | 72-Hour Kit

These specially formulated chews for dogs help calm them in stressful situations. They contain no drugs, so they are safe to give without a veterinary prescription. The active ingredients are C3 Colostrum Calming Complex, which is isolated from proteins that have a calming effect on animals; L-theanine, an amino acid that supports cognitive functions and stress reduction; and thiamine (vitamin B1), which supports the central nervous system and the relaxation response. Composure for dogs is available in chicken liver–flavor bite-sized chews, and can be used either for immediate support or daily for ongoing support. They come in a sixty-chew pouch for large dogs or a thirty-chew pouch for small dogs.

VETRI-SCIENCE COMPOSURE FELINE BITE-SIZED CHEWS
$ | Amazon.com | Go-Bag | 72-Hour Kit

These specially formulated chews for cats help calm them in stressful situations. They contain no drugs, so they are safe to give without a veterinary prescription. The active ingredients are C3 Colostrum Calming Complex, which is isolated from proteins that have a calming effect on animals; L-theanine, an amino acid that supports cognitive functions and stress reduction; and thiamine (vitamin B1), which supports the central nervous system and the relaxation response. Composure for cats comes in chicken liver–flavor bite-sized chews, and can be used either for immediate support or daily for ongoing support. They come in a thirty-chew pouch.

ANXIETY WRAP FOR DOGS

$$ | DrsFosterSmith.com | Go-Bag | 72-Hour Kit

To keep your dog more relaxed, this dog wrap applies gentle all-over pressure, like a hug, that reduces anxiety. During a disaster this can be critical for everyone's safety. Anxiety Wrap was also designed to target specific acupressure points. It's made of a lightweight fabric, and has a front head opening so you can easily slip it over your pet's head. It also includes front leg openings, a belly flap, two rear leg straps, and an open rear end for elimination. It comes in eleven sizes. There have been scientific studies done that find this wrap to be effective.

THUNDERSHIRT FOR CATS

$$ | DrsFosterSmith.com | Go-Bag | 72-Hour Kit

With the same anxiety-reducing properties as the Anxiety Wrap for dogs, this product comes in three cat sizes and is designed to apply pressure that helps reduce the anxiety experienced by felines during a stress-inducing event. The use of gentle pressure has not been studied in cats the way it has in dogs, but there are anecdotal reports that it has a similar effect. A cat should become accustomed to wearing the Thundershirt before a stressful event. The website that sells these shirts includes instructions on how to do that, under the tab "More Information."

What to Include in Your Pet's Go-Bag

PAPERWORK (IN A WATERPROOF POUCH)

- Feeding instructions
- Immunization records (required by many pet shelters)
- Pet name and detailed identification information
- Recent photo of a family member with the pet

HEALTH AND HYGIENE ITEMS

- 3 to 7 days' worth of canned (pop-top) and/or dry food (be sure to rotate the food every two months)
- Animal first aid kit
- Any medications with instructions
- Bottled water and water bowl
- Feeding bowl
- Household chlorine bleach for sanitizing
- Kitty litter and litter box with scooper for cats
- Liquid dish soap and disinfectant
- Newspaper or other material for animal elimination
- Parasite control medications
- Plastic bags for disposing of used cat litter or dog feces

TOYS AND BEDDING

- Any familiar toys, or a sleeping pad or blanket that looks and smells familiar, will help calm and occupy the animal.

TRAVEL GEAR

- Blanket (for scooping up a fearful pet)
- Extra collar or harness, and an extra leash
- Traveling bag, crate, or sturdy carrier for each pet

Learn More The ASPCA (ASPCA.org), the Red Cross (RedCross.org), and FEMA (Ready.gov) all have sections of their websites devoted to disaster preparedness for animals.

On Alert

- Never leave a pet outside during a storm or any disaster, either tied up on a leash or loose.
- Never leave your pets behind. Pets are not allowed at all shelters, so have a place for them planned in advance, such as a local motel or animal shelter where animals are allowed, or the home of a friend or relative. If you have more than one pet, be aware that you may need to board them in separate locations, and pack for each animal accordingly.
- Make sure you have a current photo of every pet for identification purposes. Put it with the important papers in your own Go-Bag.

CHAPTER 13

Auto Preparedness

CHAPTER 13

Auto Preparedness

Think fast: How much gas is in your car right now? A prepared person makes sure it's never less than half a tank, because in a disaster you may not be able to buy gas—which means you might not be able to evacuate as quickly as you should.

A well-prepared car is one that will help you survive the kinds of threats you might face while on the road, from a splatter of road insects to difficult road conditions during an evacuation, with your family and pets sharing the ride. A well-prepared car is always well maintained. Keeping your car ready means routinely checking these items:

- Air conditioner fluid
- Antifreeze
- Battery
- Brake fluid
- Brakes
- Defroster
- Emergency flashers
- Exhaust
- Fuel
- Heater
- Ignition
- Oil
- Radiator
- Tire pressure and wear
- Windshield wiper fluid

A well-prepared car is also well stocked with emergency items. Keep your Car Kit in your car, rather than in the garage, because if disaster strikes when you're not at home, you'll still have it with you. In addition to the items in your Car Kit (which you'll find at the end of this chapter), we are presuming you will take your Go-Bag (see chapter 1), and that if you live someplace where it's very cold, that you have boots and warms socks as well.

Everyone's car has a different capacity for storage, so for drivers of sports cars, the suggestions for emergency supplies are excessive, and for drivers of full-sized SUVs and trucks, perhaps not expansive enough. Keeping this in mind, do your best to fit the most critical supplies in your vehicle, and in an evacuation, make sure your Go-Bag fits along with your Car Kit. Remember, use car common sense; your kit will not look exactly like someone else's. We believe it's useful to think carefully about each item, because it gets you deeper into the necessary mindset for disaster preparedness.

Supplies You'll Need: Car

COLEMAN CABLE 08660 HEAVY-DUTY 4-GAUGE AUTO BATTERY BOOSTER CABLES WITH POLAR GLO-WATT CLAMPS
$$$ | Amazon.com | Car Kit

These auto jumper cables are twenty feet long, with four-gauge wire, and have been well reviewed for being heavy and consistently reliable. They are engineered to be effective in both hot and cold weather, according to the manufacturer. The clamps that attach to the positive and negative posts on the battery have triple polarity identification: they're color-coded, indent stamped, and marked with glow-in-the-dark labels, making it less likely for you to attach the cables incorrectly.

RESCUE TAPE SELF-FUSING SILICONE
$ | Amazon.com | Car Kit

Definitely put this in your Car Kit. A step up from the garden variety electrical tape, this stuff makes an instant seal that's waterproof and air tight and can resist heat up to 500 degrees Fahrenheit, yet remain flexible in cold down to minus 85 degrees Fahrenheit. That's going to be very handy for quick repair of hoses and gaskets. The roll is twelve feet long and the tape is one inch wide and black.

BUCKET BOSS 06009 JUMPER CABLE BAG
$ | Amazon.com | Car Kit

This bag is not an essential, but in a road emergency, you don't want to be untangling your jumper cables or fumbling around with a broken zipper on a no-frills cable sack. Gloves and other necessities fit inside with the cables, making cable storage more organized. The bag is small enough to stow inside your spare wheel.

SMITTYBILT CC330 RECOVERY STRAP
$$ | Amazon.com | Car Kit

This cheerfully yellow towing strap could make the difference between your car being in the creek or being back on the road. It is three inches wide and thirty feet long, and it's rated for thirty thousand pounds, so it will hold up when it is strapped around the bumper or other part of a car that needs towing. This item is good for many disaster situations, including one in which you might come to the aid of another driver.

FIRST ALERT AUTO FIRE EXTINGUISHER
$ | Amazon.com | Car Kit

Even if an aerosol fire extinguisher fits better in your car, don't bother. Ratings professionals report they are not reliable. This extinguisher and mounting bracket is specifically for extinguishing car fires—oil, grease, gasoline, and electrical fires. It comes with a five-year warranty and is UL rated. (See chapter 7 for more information about fire extinguishers.)

AAA 4342AAA EMERGENCY WARNING TRIANGLE
$ | Amazon.com | Car Kit

Warning reflectors are essential in your Car Kit. This bright-orange, seventeen-inch emergency warning triangle is effective on both sides, and in daylight and darkness. It is U.S. Department of Transportation (DOT) approved, and is easy to set up. Its nonskid base is designed to withstand mild wind gusts.

FLAREALERT 9-1-1 LED EMERGENCY BEACON FLARES WITH STORAGE BAG
$$ | Amazon.com | Car Kit

This type of nonburning flare lasts longer than traditional flares in an emergency, and is brighter and more easily seen. Customers rate the magnetic feature as very positive, because the flare attaches (via magnet) to the car for greater visibility. Although it is called a flare, it is really a battery-powered beacon light with two brightness settings, drawing attention that lets other drivers and rescue teams know you are having a roadside emergency. Each beacon requires three AAA batteries (not included).

NO-SPILL 1405 2.5-GALLON POLY GAS CAN

$$ | Amazon.com | Car Kit

Users say this gas can actually does what the name suggests: no-spill trans-
fer of fuel. You can buy this product in larger and smaller sizes to suit your
car size. Bigger trucks with bigger gas tanks might need at least a five-gallon
capacity, while this smaller can will be enough to get a sedan back on the road
to the gas station. This product meets the manufacturing criteria set by the
EPA and by the California Air Resources Board (CARB).

HOPKINS FLOTOOL 10705 GIANT FUNNEL

$ | Amazon.com | Car Kit

For more safely pouring fuel and oil in an engine, this funnel will help pre-
vent spills and will filter particles out of the engine. Too large for use in a tiny
car engine, this funnel is more likely to be useful in trucks, heavy equipment,
or sport utility vehicles, most of which have truck engines. Owners of smaller
cars can use (and keep in their Car Kits) the small, disposable paper funnels
that are available in auto parts stores.

GARMIN NÜVI 3490LMT 4.3-INCH PORTABLE GPS NAVIGATOR
WITH LIFETIME MAPS AND TRAFFIC

$$$$ | Amazon.com | Car Kit

Maps are much cheaper and GPS systems are numerous, but if you sim-
ply must have a sophisticated technological navigation system that is also
reliable, this is the one to get. It has earned the highest ratings both from
Consumer Reports and from the editors at *PC Magazine*. The Garmin has
reportedly demonstrated the best performance among GPS devices now on
the market, and can help you evacuate in the right direction.

MEDIQUE 40061 FIRST AID KIT
$ | Amazon.com | Car Kit | Work Kit

For those who do not wish to build their own first aid kit, this OSHA-approved kit is designed for on-the-go use and is fine for your car. It contains sixty-one items—including the standard bandages, pain relievers, and ointments a kit requires—in a plastic carrying case, and will fit into most vehicles without taking up too much space. It is also recommended for use in the workplace, so you can buy one for the car and one for the job.

EMERGENCY ZONE BRAND EMERGENCY PONCHO
$ | Amazon.com | Car Kit

Inexpensive and well-reviewed, this emergency rain poncho takes up very little space (it can fit in a back pocket) and is emergency wear you will be glad to have in your Car Kit. If you are ever forced to change a tire in wet weather, for example, this little plastic garment gem will keep you from getting soaked—which could lead to hypothermia in a worst-case scenario. You might also use it as a wind-breaker.

RESTOP 2 DISPOSABLE HUMAN WASTE BAGS
$$$ | Amazon.com | 72-Hour Kit | Car Kit

Garbage bags are cheaper, yes. But each of these bags is specifically designed (and contains chemicals agents) to contain and neutralize urine, and can be used anywhere, with or without a bucket or folding toilet. They come in a package of twenty-four packets. Each packet contains one bag, plus toilet paper and hand sanitizer. You can put a few in your Car Kit and use the others in your 72-Hour Kit at home.

AAA WARRIOR ROAD KIT
$$$ | Amazon.com | Car Kit

If you don't want to assemble your own Car Kit, this one is recommend by *Popular Mechanics* as the best all-purpose road-emergency kit on the market. This seventy-seven-piece kit includes the basic first aid supplies as well as hand tools that are considered essential for the typical kinds of tasks one must perform in roadside emergencies. Among other essential items bundled into this kit are jumper cables, making this a good second choice to thinking it through yourself.

What to Put in Your Car Kit

- Adjustable wrench
- Batteries
- Boots, wool socks
- Car fluids
- Duct and rescue tape
- Emergency road reflectors
- Fire extinguisher
- First aid kit
- Flashlights
- Flat-head folding shovel
- Funnel
- Fuses
- Gloves
- Jumper cables, 8 to 12 feet
- Knife
- Maps/GPS
- Phillips screwdriver
- Rags
- Rain gear

- Road flares
- Screwdriver
- Small gas can
- Space blanket
- Tire chains
- Toilet paper and plastic bags
- Tow strap
- WD-40
- Wool blanket

On Alert

- If you don't already have snow tires, carry tire chains in the winter if there is any possibility you may be driving in snow. Practice putting on tire chains in your driveway; you will then be prepared to apply them quickly in freezing weather and under dangerous roadside conditions.
- In your Car Kit, bring along the essential fuses and auto fluids recommended in your car manual or by your mechanic.
- Flashlights and batteries are essential (see chapter 6); make sure you get a waterproof flashlight. A headlamp is useful for changing tires at night and for any other hands-free work you might be faced with.

CHAPTER 14

Workplace Preparedness

CHAPTER 14

Workplace
Preparedness

E xperts predict that if San Francisco were to have a major earthquake, people evacuating in the heart of the Financial District could be faced with unexpected barriers, such as broken window glass piled as high as twelve feet in some areas. It's hard to imagine; even one foot of broken glass would present a daunting obstacle. Experts in urban-disaster preparedness have to look at all the possibilities.

We need to look at our workplaces, wherever they may be, and plan for every disaster we can imagine. (The Red Cross catalogs twenty-two emergency scenarios, in case your imagination needs some prompting; you can see them all at RedCross.org.) Keep a personal emergency kit and a pair of socks and good walking shoes at work. Encourage your employer to store water and other emergency supplies for the staff. Find out your workplace emergency plans for all events, from storms to bomb threats. Familiarize yourself with these emergency plans and get to know your building and its every egress.

If a tornado or hurricane threatens, be aware of your exposure to windows. You don't need to be an architect to notice which direction the building faces and in what kinds of weather the windows might be vulnerable. If you live in earthquake country, identify the heaviest piece of furniture near your work station that might serve as a place to duck under for the duration of a tremor or quake, during which falling objects present the greatest danger.

Just as professionals suggest you do in your neighborhood, make note of the most vulnerable people at your workplace and become more aware of their particular needs. Anyone diabetic? Elderly or frail? Are there those whose physical limitations—including obesity—might mean they require physical assistance in an evacuation? The people who occupy your floor or building may likely include those with whom you never mix in the course of your workday, but with whom you might find yourself closely aligned in an emergency. When Hollywood films depict an unlikely assortment of characters together in a stuck elevator, it's not an implausible plot device—emergencies are often as random as the impromptu "team" that forms to survive them.

People who have completed emergency skills training programs, such as CERT, can make all the difference when a disaster strikes away from home. Many workplaces have a designated emergency floor director, the go-to person who is familiar with emergency protocols, who makes sure emergency supplies are rotated and fresh, and who makes sure the stairs and not the elevator are used in emergency evacuations.

The Work Kit Your Work Kit should contain some personal items and some things for the workplace as a whole. What you can keep at your workplace will be determined in part by where you work—a restaurant, business office, retail store, or factory. Aside from whatever personal items you can manage to store in your kit at work, your workplace as a whole should have a pared-down version of your 72-Hour Kit. For personal items, consider nutrition bars, coins and single dollars, water, an extra pair of glasses or contact lenses, medications, and identification, in case the disaster keeps you from home for a while. Make sure you have an emergency cell phone charger, too. It may make you the most popular person at work.

Things Your Employer Needs to Know

Bring your employer's attention to the following items and personnel tasks for emergencies, especially if you work in a tall building with elevators.

- CERT training for employees
- Designated emergency manager for your floor
- Designated people to carry the evacuation chair
- Designated people to use the fire hose in emergencies
- Designated person to check the water supply and its shelf life
- Designated person to make sure first aid supplies are maintained
- Large-capacity first aid kit located in an easily accessed area (and perhaps an automated external defibrillator—AED)
- Occupant-use fire hoses
- Red Cross training for employees
- Regular disaster drills
- Wall-mounted evacuation chair for those in wheelchairs who cannot use the stairs
- Water stored just for emergencies

Wait It Out The odds of being stuck in an elevator are actually fairly high, but the odds of that elevator dropping off its cable and plunging to the ground are very low. The last time an elevator dropped was in the 1940s, when a U.S. Army plane crashed into the Empire State Building; the woman in the elevator survived the fall. Never attempt to get out of a stuck elevator. Wait for help.

Supplies You'll Need: Workplace

ACME THUNDERER METAL WHISTLE
$ | GunDogSupply.com | Work Kit

There are cheaper plastic whistles, but in an emergency, a full-throated screaming device will better help rescuers locate you. This nickel-plated brass whistle, developed in England for the London police, should do the job.

3M 1860 HEALTHCARE PARTICULATE RESPIRATOR AND SURGICAL MASK
$ | Amazon.com | Work Kit

In a debris-filled atmosphere, you can breathe safely with a mask. Any mask you buy should have a N95 rating, as this one does. (That means it filters at least 95 percent of airborne particles, but it is not resistant to oil.) It's FDA-cleared, latex-free, has a molded cone design, and is resistant to fluids and splashes. It offers bacterial filtration efficiency greater than 99 percent. This mask fits a wide range of face sizes.

ADVENTURE MEDICAL KITS ULTRALIGHT AND WATERTIGHT
$$ | Amazon.com | Work Kit

Since water (even from sprinkler systems that go off when there's not fire) can play such a large role in workplace emergencies, we suggest this water-proof kit used by campers and hikers. What makes this kit distinct among well-supplied standard first aid kits is that the vulnerable supplies inside it will remain useable after being dropped in water. It contains wound care items, bandages, alcohol swabs, wraps, and a few basic medications.

MSA SAFETY WORKS 818068 HARD HAT

$ | Amazon.com | Work Kit

You can use this bright-yellow hat not only to identify you as the emergency floor manager at your workplace (unless hard hats are part of your workplace attire, in which case you will have to get a different color), but also to protect your head from falling debris during evacuation. The shell is made of polyethylene, and the adjustable crown suspension straps inside are nylon.

ADVENTURE MEDICAL KITS TRAVEL DUCT TAPE 2-PACK

$ | Amazon.com | Work Kit

Make room in your Work Kit for all-purpose duct tape. This tape comes in two four-foot rolls. They don't have a cardboard core, so you can flatten them to fit into a smaller kit. Duct tape can be used to make a space-blanket pouch or bag (see chapter 7) or to create a shelter in place (see chapter 4), or in many other emergency scenarios.

ZOLL AED+ BUSINESS VP

$$$$$ | Amazon.com

This automated external defibrillator (AED) is, of course, your employer's purchase! Lifesaving professionals report that these devices have saved lives that might have been lost if the victim would have waited for EMTs or paramedics to arrive on the scene. The device is designed so that almost anyone can apply electrical charges to the chest of a person who is having a heart attack, keeping the person alive until professional medical help arrives. It comes with a metal wall cabinet and the required batteries.

---

20 PERSON ULTIMATE DELUXE SURVIVAL KIT (SK20R)

$$$$ | QuakeKare.com

Another purchase to be made by your employer (or you, if you employ others), this 72-Hour Kit includes a smartphone charger and a five-gallon plastic bucket toilet with all the necessary hygiene-related tools, as well as a standard hand-crank emergency radio, a flashlight, and a standard first aid kit for twenty people. You can purchase a larger kit if your workplace has more people, or several of these kits.

On Alert

- For a better chance of survival, network with others at your workplace: talk about disaster preparedness, create buddy systems, and learn emergency response skills together.
- In an earthquake, get under heavy furniture or crouch down at the base of a structural (not a cubicle) wall, and lock your intertwined fingers at the back of your neck.
- Keep a pair of good walking shoes or boots and a pair of socks with your Work Kit.

APPENDICES

Guidelines for Food

PLAN A: FOOD STORAGE GUIDELINES

If you plan to use Food Plan A, use the Federal Emergency Management Agency (FEMA)–Red Cross chart below to group emergency foods by storage duration. (Packages should always be stored unopened.)

USE WITHIN 6 MONTHS	USE WITHIN 1 YEAR OR BEFORE DATE ON PACKAGE	INDEFINITE SHELF LIFE UNDER PROPER CONDITIONS
Dried fruit	Canned fruits, fruit juices, and vegetables	Baking powder
Dry, crispy crackers	Canned meat and vegetable soups	Bouillon products
Freeze-dried potatoes	Canned nuts	Dried corn
Powdered milk (boxed)	Hard candies	Dry pasta
	Jelly	Instant coffee, tea, and cocoa
	Peanut butter	Noncarbonated soft drinks
	Ready-to-eat cereals and uncooked instant cereals	Powdered milk (in nitrogen-packed cans)
	Vitamins	Salt
		Soybeans
		Vegetable oils
		Wheat
		White rice

PLAN B: FOOD SUPPLIES

Planning for an emergency can be both overwhelming and time consuming. Food Plan B provides a direct path to ensuring you and your loved ones' nutritional needs are taken care of when disaster strikes. Food Plan B kits come in bulk quantities and will require some storage space. These easily transportable options offer simple-to-prepare meals, long-term storage, and a kit suitable for almost every budget.

COMPANY	VENDOR	SHELF LIFE	SERVINGS	CONTAINER	COST (APPROXIMATE)
Chef's Banquet	Costco.com	20 years	11,352	Pallet of 36 six-gallon buckets containing individual pouches	$4,000
Augason Farms Meal Pack (gluten-free)	SurvivalFood.com	30 years	1,185	18 no. 10 cans, packed in three boxes	$270
Deluxe Survivor Variety Food Storage (vegetarian)	Costco.com	20 years	700	2 six-gallon buckets containing individual food pouches	$200
Chef's Banquet ARK I and II Ultimate Entrée Combo	Costco.com	20 years	638	2 six-gallon buckets containing individual pouches	$200
Chef's Banquet Emergency Food Storage ARK	Costco.com	Up to 20 years	330	sixteen-gallon bucket containing individual pouches	$100

Additional Resources

Government Resources for Preparedness and Emergencies

CENTERS FOR DISEASE CONTROL AND PREVENTION (CDC)

This organization looks at disasters from a public health perspective, and offers the general public helpful advice about stopping the spread of infectious disease during disasters.

CDC.gov | (800) 232-4636

DEPARTMENT OF HOMELAND SECURITY

This website is a resource for all kinds of protective efforts nationwide. If you search for disaster preparedness information, it links you to Ready.gov.

DHS.gov | (202) 282-8000

DISABILITY.GOV

This site enables you to zero in on your state and region for preparedness-related information tailored specifically for those with disabilities and the people who care for them.

Disability.gov

FEDERAL EMERGENCY MANAGEMENT AGENCY (FEMA)

FEMA is an enormous administrative arm that oversees the orderly response to disasters of all kinds. For citizen preparedness, it links to Ready.gov.
FEMA.gov | (202) 646-2500

FLU.GOV

This site is geared toward organizations rather than individuals, and offers education materials for community groups to use in helping inform citizens about what to do to stop the spread of the flu.
Flu.gov

HOMELESSNESS RESOURCE CENTER

Disaster preparedness for the homeless is taught by mental health providers through this website, a project of the U.S. Substance Abuse and Mental Health Services Administration.
Homeless.SAMHSA.gov

NATIONAL WEATHER SERVICE

The National Oceanic and Atmospheric Administration (NOAA) produces this website and the emergency broadcast of weather advisories. On the website you can find maps for all types of weather data, and sign up to be a volunteer weather reporter.
Weather.gov

NUCLEAR REGULATORY COMMISSION

This website gives information about what to do in case of a nuclear plant meltdown or a radioactive "dirty" bomb detonation. Those within ten miles of a nuclear plant are considered inside the emergency zone of a nuclear disaster.
NRC.gov | (800) 368-5642

OCCUPATIONAL SAFETY AND HEALTH ADMINISTRATION (OSHA)

OSHA is charged with keeping workers in America safe, and has a lot of information about how to protect against everything from caustic chemicals to natural disasters. The information is geared to the workplace environment, but is just as valid for your home.

OSHA.gov | *(800) 321-6742*

READY.GOV

This is a user-friendly public service of FEMA and the number-one go-to site for preparing for disasters. You can drill down to fine details here.

Ready.gov | *(800) 462-7585*

UNITED STATES GEOLOGICAL SURVEY (USGS)

The USGS site gives detailed information about earth-related threats, with in-depth coverage of earthquake geology. You can sign up for local earthquake alerts to be sent to you by e-mail.

USGS.gov | *(703) 648-5953*

Non-Government Resources for Preparedness and Emergencies

AMERICAN RED CROSS

A multifaceted relief organization, the American Red Cross responds to about seventy thousand disasters annually, trains citizens in first aid and CPR, and produces health and safety educational materials, including a website.

RedCross.org | *(800) 733-2767*

AMERICAN SOCIETY FOR THE PREVENTION OF CRUELTY TO ANIMALS (ASPCA)

The ASPCA is a nonprofit animal advocacy group whose website contains a section on disaster preparedness for animal owners. The website offers free "animal inside" alert stickers for your home.
ASPCA.org | (212) 876-7700

AMERICAN VETERINARY MEDICAL ASSOCIATION (AVMA)

The AVMA website provides information for the public, including an emergency preparedness section and information on caring for your pet's health.
AVMA.org | (800) 248-2862

DISASTERSAFETY.ORG

This is a service of the Insurance Institute for Business and Home Safety. Their mission is "to conduct objective, scientific research to identify and promote effective actions that strengthen homes, businesses, and communities against natural disasters and other causes of loss."
DisasterSafety.org | (813) 286-3400

EQUIPPED TO SURVIVE

This is a nonprofit organization dedicated to raising awareness about emergency preparedness among the general population. The group's disclaimer is that they are not "survivalist" and do not endorse survivalist values.
Equipped.org

Emergency Services

IN EMERGENCIES THAT THREATEN YOU OR OTHERS, ALWAYS CALL 911.

Emergency services	*911*
Nationwide traveler information	*511*
Exposure to chemicals or biological agents	*911*
CDC bioterrorism preparedness and response	*(404) 639-0385*
CDC environmental health/toxic substances and disease registry	*(770) 488-7100*
Chemical Facility Anti-Terrorism Standards (CFATS) security tip line	*(877) 394-4347*
Report suspicious activity in federal buildings	*(877) 437-7411*

ACKNOWLEDGMENTS

We at Novato Press wish to thank the following individuals for their generous contributions of expertise in developing the factual content of this book. Any errors are ours alone.

Steve Brassfield
Battalion Chief
Napa City Fire Department
Napa, California

Carl Johnson
Fire Captain and CERT Instructor
Napa City Fire Department
Napa, California

Darren Drake
Fire Marshal
Napa City Fire Department
Napa, California

Darryl Madden
Department of Homeland Security
Ready.gov
Washington, D.C.

Patrick Gilleran
Senior Engineer
Administrative Office of the Court
State of California

INDEX

Printed in May 2022
by Rotomail Italia S.p.A., Vignate (MI) - Italy